Technician's Handbook of Plastics

Technician's Handbook of Plastics

Peter A. Grandilli

VNR VAN NOSTRAND REINHOLD COMPANY
NEW YORK CINCINNATI ATLANTA DALLAS SAN FRANCISCO
LONDON TORONTO MELBOURNE

Van Nostrand Reinhold Company Regional Offices:
New York Cincinnati Atlanta Dallas San Francisco

Van Nostrand Reinhold Company International Offices:
London Toronto Melbourne

Library of Congress Catalog Card Number: 80-19379
ISBN: 0-442-23870-3

Manufactured in the United States of America

Published by Van Nostrand Reinhold Company
135 West 50th Street, New York, N.Y. 10020

Published simultaneously in Canada by Van Nostrand Reinhold Ltd.

15 14 13 12 11 10 9 8 7 6 5 4 3 2 1

Library of Congress Cataloging in Publication Data

Grandilli, Peter A
Technician's handbook of plastics.

Includes index.
1. Plastics—Handbooks, manuals, etc. 2. Polymers and polymerization—Handbooks, manuals, etc. I. Title.
TP1130.G75 668.4 80-19379
ISBN 0-442-23870-3

Preface

The primary intent of the author is to furnish the plastics technician with an on-the-job reference book that is broad in scope, yet concise and to the point. *The Technician's Handbook of Plastics* is designed to not only help with current assignments, but also to familiarize technicians with other areas of plastics technology. The added knowledge will not only aid in meeting new challenges but will also add to qualification for more prestigious and financially rewarding positions. Because of the unprecedented, broad presentation of detailed "how to," the book will also be of much assistance to plastics engineers, supervisors, and instructors as a means of quick reference.

The secondary purpose of *The Technician's Handbook of Plastics* is to present to beginners and students at all levels a comprehensive study of industrial plastics. The text provides an in-depth look at methods, materials, equipment, and techniques and is scaled to the layman's level of comprehension. Basic fundamentals, nomenclature, and even shop jargon are discussed. Those planning a career in plastics or a career associated with plastics can now better prepare themselves for the work ahead.

With the great advance in technology and widespread use of plastics has come the demand for more people trained and skilled in the many phases of this phenomenal industry. No other field offers a better opportunity for the beginner who is short of a college degree. Plastics are being used in more and more ways, and one can hardly go anywhere without seeing one of these versatile materials in use. In a modern house, plastic may be part of the wall insulation, part of the telephone, radio, and television set, and used in floor tile, wall paint, refrigerator components, furniture, mixer housings, shower stalls, tabletops, and many more materials and furnishings too numerous to mention. Even the books one reads may be coated with a plastic material to improve the strength and water resistance of its binding and jacket. Away from the home, plastics are used in the making of automobiles, street lights, buses, boats, aircraft, radomes, space vehicles, and thousands of other objects.

I would like to thank all of the materials and equipment manufacturers who provided pictures and drawings for the text. In particular, I would like to express my gratitude to Monsanto Co., Eastman Chemical Products, Uniglass Industries, CY/RO Industries, Uniloy, and Durez for their technical contributions. I also wish to thank Robert R. Arnold, photographer, and William Wood, artist. Last but not least, I would like to thank my lovely daughter Jeanne for typing the manuscript. Her secretarial skill, hard work, and organizational direction were major factors in the preparation of this book.

PETER A. GRANDILLI

Contents

	Preface	v
1.	Introduction	1
2.	Basic Chemistry and Properties of Plastic Materials, by Harry Raech, Jr.	8
3.	Injection Molding, Blow Molding, Extrusion	34
4.	Acrylic Fabrication	81
5.	Cellulosic Fabrication	86
6.	Miscellaneous Thermoplastic Processes	93
7.	Curing Systems for Epoxies and Polyesters	98
8.	Reinforcements for Thermosets	103
9.	Compression and Transfer Molding	111
10.	Hand Layup	134
11.	Miscellaneous Thermoset Processes	147
12.	Plastics in Electronics	155
13.	Thermoset Labs	181
14.	Physical Tests for Plastics	191
15.	Mechanical Tests for Plastics	198
16.	Reference Tables and Guides	207
	Glossary	235
	Index	243

Technician's Handbook of Plastics

1 Introduction

Before the advent of synthetic plastics, most forms of construction and production were dependent upon the use of natural materials such as metal, wood, rubber, tar, and various minerals. These materials are still very much in evidence today, but in many instances they have been replaced by the new wonder materials called plastics. Through the magic of the chemist, we now have plastics that can meet almost every engineering requirement.

The first man-made plastic was nitrocellulose. It was developed by chemists in both America and Europe who had been trying to find new uses for cellulose, the basic substance of wood and cotton. They finally succeeded in synthesizing the new chemical by treating cotton fibers with nitric acid. In 1870, a young printer named John Wesley Hyatt and his brother added camphor to nitrocellulose to form the first commercially successful plastic, which they called *celluloid.* The new material was used to make combs, brushes, photographic film, lacquers, and early automobile windows and safety glass. But celluloid had two inherent weaknesses that have severely curtailed its use in modern times: it is highly flammable, and in the form of clear film it yellows upon aging.

The second plastic to appear on the market was called *Bakelite.* It was developed by Dr. Leo Baekeland in 1909. Dr. Baekeland had been searching for a synthetic coating material to replace shellac, which is derived from secretions of the lac bug and other insects. He began experimenting with two chemicals, phenol and formaldehyde, and after many experiments, combined the two with catalysts to form the new material. In addition to being used in varnishes and lacquers, Bakelite was also used for molding, casting, adhesive bonding, and electrical insulation.

Bakelite today is called phenolic and has developed into one of our most popular synthetic plastics.

WHAT ARE PLASTICS?

Most plastics are man-made materials with the ability to flow, take shape, and solidify. They are made from chemicals extracted from substances such as petroleum, coal, air, water, and agricultural by-products. The production sequence starts with the manufacturer of raw materials who combines the various chemicals to form meltable solids and syrupy liquids called *resins*. Next, plastics fabricators purchase the raw materials and convert them into end products through the use of appropriate methods and equipment.

Unlike metals, plastics are light in weight, pleasant to the touch, and easy to form. Complicated shapes are easily reproduced by molding, casting, or laminating. Because they do not allow electricity to pass through them, they are used extensively in dielectric applications. They are also poor conductors of temperature, which accounts for their use as pot handles, electric iron handles, refrigerator housings, and Thermos bottles.

Plastics are very versatile materials. Some have great optical clarity, and most can be colored in a wide range of colors and shades. One plastic, acrylic, is even clearer than glass. Because the dyes for colored plastics are mixed with the raw materials, the completed parts need never be painted. Plastic surfaces are also free from the atmospheric corrosion that is prevalent with some metals. The light weight of plastics is still another advantage over metals. Some, such as polypropylene, are lighter than water and therefore float on water.

Lightweight Structures

The ability of plastics to be foamed has led to a completely new concept of lightweight construction. Rigid urethane foams are already finding wide use in furniture and other structural applications. The strength of a rigid urethane foam panel weighing 8 lb per cu ft is something to behold, expecially when one considers the fact that water weighs 62.4 lb per cu ft.

Another extraordinary development in lightweight structure is the combining of plastic binders and adhesives with *honeycomb* and facing materials to form panels with outstanding weight-to-strength ratios. Bees may have invented honeycomb structures, but it took human ingenuity to turn an empty honeycomb to commercial advantage. Commercial honeycomb is 97% air and 3% metal or fiber by volume, and to build an industry out of something that is 97% air is a rare accomplishment. Honeycomb applications have invaded such basic fields as transportation, architecture, illumination, aircraft and guided missiles, electronics, materials handling, home appliances, and instrumentation.

The world of construction is certainly looking to working with high-strength honeycomb panels and lightweight plastics like rigid urethane foam as a promise of tomorrow. For example, with ever-decreasing supplies of oil and other forms of energy has come the demand for more and more lightweight vehicles, the construction of which plastic will make possible. Indeed, both the engineer and the architect are in constant search for materials that will improve economy, weight-to-strength ratio, and ease of fabrication with no sacrifice in performance.

Metallized Plastics. Many plastics can be coated with a thin metal film to change their appearance or meet a functional requirement. ABS, in particular, is very adaptable to the film-depositing process.

Bright, glossy, chromium-coated plastics are especially popular. They are difficult to distinguish from solid metal unless lifted by hand, at which time the much lighter plastic is very discernable.

Plumbing fixtures and toys are among the more popular applications of metallized plastics.

Human Implantation. An important breakthrough in surgery has been the use of plastics as implantations in the human body. Strong, tough, nonallergenic plastics are finding increased use as replacements for defective blood vessels and other damaged parts of the body. The softer silicone rubbers are also used for implantation, but mostly for cosmetic purposes.

Mass Production. Plastic products are now being produced more quickly and in larger volumes, thanks to the tremendous strides made in injection, compression, and transfer molding. Improved mechanization of automatic and semiautomatic machines plus improved molding materials have brought about shorter cure cycles and a reduction in handling operations.

Chemical Resistance. Many plastics can be used in specialized chemical-resistant applications. With few exceptions (such as the fluo-

rocarbons), most plastic materials are attacked by one or more chemicals. But, though vulnerable to some chemicals, a specific plastic can be highly resistant to several others. By careful material selection, lightweight, unbreakable plastics can be used to replace glass and stainless steel as chemical containers. In many instances, plastics have replaced stainless steel as industrial ducting used to carry away chemical vapors.

Advantages of Metals.

Metals are generally superior to plastics in strength and heat resistance. Though some plastics can withstand high temperatures for short-term exposures, none can match some metals for continuous exposure to high temperatures. Most structures and manufactured articles, however, are not required to withstand high temperatures during normal use.

At this point, it can be observed that metals and other materials have remained readily available partly due to the incursion of plastics into their markets. Although it is true that some of the raw materials for making plastics can also become scarce, unlike metals, plastics can be made from many different, easily obtainable materials, even corncobs and oat hulls. The chemist is ever ready to wave a magic wand and bring us more synthetic plastics made from a new assortment of raw materials, if necessary. And the engineer is happy to have all types of materials available, especially when they can be combined for ultimate beauty, economy, and function.

Types of Plastics

There are two main groups of plastics: thermoplastics and thermosets.

Thermoplastics are materials that remain permanently fusible. This means that they will melt when exposed to sufficient heat. For each thermoplastic, there is a specific temperature at which the material will start to distort. This is known as the *heat distortion point*. Most raw materials used in the production of thermoplastic products are meltable solids in the form of granules and powders such as those used for injection molding and extrusion. They are simply heated and melted into specific shapes. Defective parts can be ground up and reused. Though thermoplastics melt readily, they do have a high melt viscosity and require pressure in addition to heat to force them into molds. The shape of the mold being used dictates the shape of the part that is being produced, and resolidification takes place when the mold is cooled below the heat distortion point of the material.

Thermoset raw materials are supplied in an uncured or partially cured state and are fully cured during the fabrication operation by the action of a catalyst or other curing agent. Some are catalyzed by the manufacturer, and some by the fabricator. All require a form of heat to initiate the chemical change. The heat can come from an oven, press, lamp, or, in some cases, from just the curing agent alone as it reacts with the resin, or with the promoter in the resin. Once cured, thermosets become permanently infusible, which means they cannot be melted down like thermoplastics.

Another dissimilarity is in the solubility of thermosets. Organic solvents will dissolve many thermoplastics but have little or no effect on most thermosets. Thermosets also have harder surfaces than most thermoplastics and are used primarily in reinforced plastics. Inherently, they are somewhat brittle but are combined with reinforcements such as fiberglass to form composites of great strength, some even surpassing metals in a weight-to-strength ratio.

Thermoset Materials

Raw Materials. Thermoset raw materials for processing into end products are available as liquid resins, soluble solid resins, coated fabrics and filaments, and molding compounds. Molding compounds are made in two forms: (1) dry, partially cured powders, granules, and chopped fabrics and filaments, and (2) soft and/or pliable bulk mixes.

B-Stage. The various stages of cure of a catalyzed thermoset resin are known as *A-stage, B-stage,* and *C-stage.* Minus the chemical explanation, A-stage resin is uncured, B-stage is partially cured, and C-stage is fully cured.

Many molding compounds and laminating fabrics are processed while in the B-stage. The materials manufacturer partially cures (advances) the resin in the materials according to order specifications. The processor, upon receipt of the material, stores the resin at the appropriate temperature until ready to use it. Some B-stage materials are stored at room temperature, but most must be refrigerated. With some thermoset systems, the chemical reaction cannot be stopped at the B-stage once it has started. Many systems that require an elevated temperature cure can be B-staged, most room-temperature curing systems cannot. In general, the more advanced the material, the greater the pressure requirement. But, the greater the pressure, the denser and stronger the part. B-staging, however, is not always feasible. The designer must know when to specify B-stage and how far to advance the resin. Some highly advanced molding compounds have limited flow and can only be used for shallow parts.

Exotherm

Many thermoset mixes give off heat during cure. This phenomenon is known as *exotherm*. Some room-temperature-curing polyesters and epoxies will exotherm severely if processed incorrectly. For example, if too much MEKP (catalyst) is added to a polyester that contains cobalt naphthenate (promoter), the mix can get hot enough to smoke and even catch fire.

Exotherm can be a help or a hindrance, depending on the application. Wet polyester layups can be processed at room temperature, using only exothermic heat to achieve cure. This is particularly advantageous when the layup is too large for available oven space. While exotherm can be an advantage to the laminator, it can be a detriment to processors involved in casting and potting. Only a few thermosets— silicone rubbers and certain urethanes—can be cast in large volumes without exotherming. Epoxy-polyamides (such as Versamids*) can be cast in large volumes because of the low heat emission during exotherm. Casting with some room-temperature-curing thermosets is confined to relatively shallow pourings. Large castings with these materials require several pour and cure cycles but, in the case of polyesters, shrinkage away from the mold can create gaps that will be filled by subsequent pourings, thus detracting from the appearance of the casting. Minimum catalyzation and the use of appropriate fillers can help prevent degradation of the casting due to exotherm.

Some thermoset mixes that require applied heat to initiate the curing action will also show an exothermic rise in temperature. During the pressing of a phenolic laminate, the temperature of the laminate will rise sharply when it reaches 220–238 °F. At the conclusion of the exothermic cycle, the temperature will return to normal.

---

*Trademark of General Mills Chemicals, Inc.

Pot Life. Pot life is the length of time that a catalyzed resin mix remains usable at room temperature. Beyond this period, the resin will start to advance, and the mix must be discarded. Pot life pertains primarily to freshly mixed liquid resin systems for laminating, adhesive bonding, potting, and casting.

Shelf Life. Shelf life is the length of time that B-staged materials, unmixed component materials, and frozen adhesives and potting compounds can remain in storage in a usable state. At the conclusion of the shelf life period, unmixed materials can be retested and recertified, but frozen adhesives and potting compounds are usually discarded because they are in a catalyzed state. Some B-stage fabrics can be reactivated by the manufacturer. Storage temperature of all materials is noted on the inspection tag.

Shop Life. Shop life is the length of time that a pre-preg remains usable at ambient temperature. A pre-preg that has a storage life of 6 months at 32°F might have only a 3-day shop life.

2 Basic Chemistry and Properties of Plastic Materials

Harry Raech, Jr.
Hughes Aircraft, El Segundo, California

With the exception of the fluorocarbons (plastics containing fluorine) and the silicones, all plastics are hardly more than chains of carbon atoms with hydrogen atoms attached to free locations on the carbons, with a maximum of three hydrogens to a carbon. When the plastics are *cured* or polymerized, the chains join together at the ends to form long, intertwined ropes and become solids. It is at this point that we can distinguish thermoplastics from thermosets. In the case of the thermoplastics, the ropes merely intertwine and are not interjoined. As a consequence, thermoplastics have relatively weak, energy-sensitive forces holding them together, and either heat or solvents can turn them into fluids. Thermosets, on the other hand, have interconnections joining the ropes into latticework. In exactly the same way that diagonals in a truss stiffen a bridge and allow it to carry a heavy load, the latticework of the thermosets stiffens them, making them more rigid as well as more temperature resistant and solvent resistant than thermoplastics. A price is paid, however, for, once cured, thermosets cannot be softened and reworked as can the thermoplastics. In addition, the thermosets cannot be solvent welded as can most of the thermoplastic resins.

The most common variation of the linear carbon-chain molecule described above is the six-carbon ring or *benzene ring* molecule increasingly found in both thermoplastics and thermosets. Phenolic was the first commercial resin to contain the benzene ring. In general, the ring structure confers improved heat and chemical resistance to a resin.

```
  h   h   h   h   h   h
  |   |   |   |   |   |
h—c — c — c — c — c — c—h
  |   |   |   |   |   |
  h   h   h   h   h   h
```

STRAIGHT-CHAIN MOLECULE

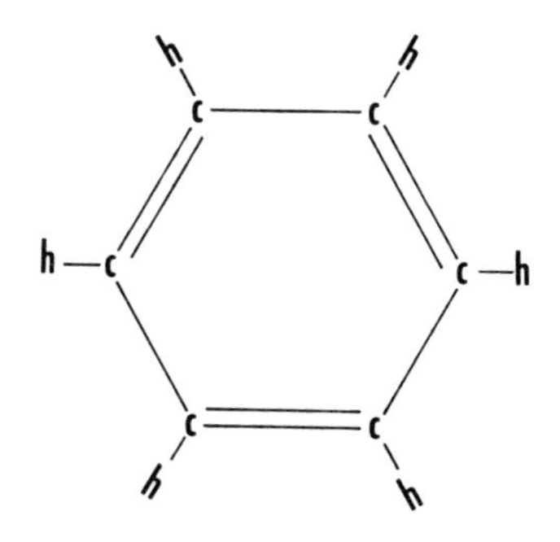

BENZENE RING MOLECULE

Basic Types of Plastic Molecules.

THERMOPLASTICS

Owing to their processing versatility, the thermoplastics are the larger branch of plastics by a considerable margin. Thermoplastics can be injection molded, extruded, thermoformed, blown, rotocast, made into coatings and adhesives, and drawn into textile fibers. They can be used in an enormous spectrum of applications from mundane textiles and tire cord to automotive and radio parts to sophisticated electronic components.

Acetal

Available since 1960, acetal is a thermoplastic in which the carbon-carbon chain has alternating oxygen atoms. It has a service temperature from -60° to 240°F. The outstanding properties of acetal, in addition to heat resistance, are high fatigue resistance, natural lubricity, high mechanical strength, and good dimensional stability. For these reasons, acetal has found an important area of application in small machine parts such as gears, levers, and bearings. Acetal is also well adapted to general-purpose components such as sprinkler housings and hardware, and the good resistance of acetal to many solvents and chemicals broadens these applications. The processing of acetal, however, has been limited to injection molding, conducted at 350–400°F.

Acetal is currently manufactured in two ways. Du Pont's Delrin is made directly from formaldehyde by a catalytic reaction. Celanese's Celcon, on the other hand, is made from the ring compound trioxane by a catalytic reaction that splits the ring to form a linear compound closely resembling that made from formaldehyde.

Acrylic

Acrylics are important transparent, light-stable resins that have been manufactured since the early 1930s, based on the work of the German chemist Rohm and that of the Englishman Hill. The major commercial application still remains that of transparencies such as glazing.

Acrylics are commercially offered as casting syrups and molding powders. The syrups find their use in casting glazing, including a very broad use as aircraft canopies that are first cast as flat sheets and then thermoformed into their final contours.

The outstanding properties of the acrylics are clarity and weather resistance, which explains their use as glazing. In addition, they possess good impact resistance and good dimensional stability. Taking advantage of these properties are typical applications such as automotive taillights, lighting fixtures, and watch crystals.

In the form of molding powder, acrylics are injection molded, extruded, and compression molded at temperatures from 350° to 550°F to form clear parts, rods and tubes.

Methyl methacrylate, the most important of the acrylics, is typically made by heating a blend of acetone cyanohydrin and concentrated sulphuric acid to produce methacrylamide sulphate. This product is then reacted with methyl alcohol and water to produce the final commercial plastic.

Methyl methacrylate is a linear carbon-chain compound having both oxygen and carbon atoms pendant at positions on the side of the chain along with the usual hydrogen atoms. It has a relatively low melting point.

Cellulosics

Cellulose Nitrate. The earliest of the synthetic plastics, cellulose nitrate has an interesting history of use as molded plastics parts, photographic film, and projectile propellent. When properly formulated with stabilizers, it is characterized by toughness, dimensional stability, and low water absorption. Its other characteristic, high flammability, has led to what is probably its major current application as an ingredient, together with nitroglycerine, of mixed-base gun powder and small rocket propellant.

A relatively brittle plastic, cellulose nitrate is usually blended with plasticizers to afford useful properties. In the past, when suitably plasticized, it was used in a variety of applications such as Ping-Pong balls and fountain-pen barrels. Other than mixed-base powder, the major present applications are films and coatings.

The molecule of cellulose nitrate is one of the most complicated among the thermoplastics. It is composed of six-sided rings having carbon in each of the positions except one, which is occupied by an oxygen atom. Pendant from this ring are carbon, hydrogen, and oxygen atoms. Alternating with the ring in the molecular chain are oxygen atoms.

Manufacture of cellulose nitrate is basically a simple process consisting of treating cellulose fibers with hot nitric acid, then washing and drying the fibers. The raw cellulose may be cotton linters (the original source), wood pulp, or paper or cotton waste.

Cellulose Acetate. Cellulose acetate has been made in the United States since World War I, although the manufacturing process was not perfected until almost 1930. It has largely replaced cellulose nitrate as photographic film owing to its low burning rate (and is now itself being threatened by newer plastics). It is also very important as a textile fiber commonly used in coat linings. Properly plasticized with up to 40% plasticizer to confer processability and toughness, cellulose acetate has been used in such diverse molded applications as appliance and tool handles, telephone handsets, and containers.

Injection molding of cellulose acetate is typically conducted with a temperature of 380°F and barrel pressures ranging from 5,000 to 8,000

psi. It is very important that the mold be heated to about 150°F to avoid defective parts.

Cellulose acetate is manufactured from the same cellulosic feedstock as cellulose nitrate. The raw cellulose is dried, then reacted with glacial acetic acid, acetic anhydride, and sulphuric acid catalyst. After hydrolysis to achieve a controlled level of chain degradation, the resin is precipitated from the chemical solution, washed, and dried. The molecule closely resembles cellulose nitrate except for the presence of the acetate group.

Ethyl Cellulose. Owing to its great toughness and flexibility and its retention of these properties at low temperature when it is suitably plasticized, ethyl cellulose is more important commercially as molded parts and coatings than cellulose nitrate. In addition, it is resistant to simple straight-chain organic solvents, although it is attacked by concentrated acids and is soluble in aromatic solvents such as benzene. It is also widely used when toughness is an important quality, as in tool handles and mallet heads, and is very popular for sporting goods such as protective helmets.

Because ethyl cellulose absorbs moisture, drying it prior to molding is very important. This may be done in a circulating oven or an infrared oven. After drying, the resin may be molded by injection, compression, or extrusion. The injection cylinder temperature is usually held at about 350°F. When compression molding, the mold is held at about 350°F and pressures up to 5,000 psi are used.

In the manufacture of ethyl cellulose, the cellulose fibers are first reacted at elevated temperature with a blend of ethyl chloride and sodium hydroxide. After the reaction is complete, the alkali cellulose is filtered from the solution and washed with ethyl alcohol. Placed in an autoclave, the alcohol-wet alkali cellulose is then exposed to ethyl chloride gas added under pressure at an elevated temperature. The final ethyl cellulose flake is then precipitated, washed, and dried.

Fluorocarbons

Polytetrafluorethylene. Polytetrafluorethylene (TFE) was the earliest of the fluorocarbons. Called Teflon in the United States, it was developed in 1938 but not brought into full commercial manufacture until 1950. Teflon is a waxy resin of the general appearance of polyethylene. However, it has heat resistance up to 550°F, remarkable lubricity, and is attacked only by sodium and other hot alkali metals. These properties have led to its usage in a spectrum of chemical- and wear-resistant applications. For example, some typical applications of TFE include bearings and bushings, bearing pads, ball-valve seats, O-rings, gaskets and packing rings, and electrical insulators.

TFE is not molded as are most resins but is treated like a ceramic. First, the material, which is supplied as a powder, is preformed into a desired shape under pressures up to 20,000 psi. This may be done in a metal tool or may be done under fluid pressure in a hydroclave. Following this, the preform is sintered in an oven to form the final part, at temperatures up to 700°F.

The Teflon molecule resembles that of polyethylene, but with the hydrogen atoms replaced by those of fluorine. It is a straight-carbon chain with two fluorine atoms attached to each carbon.

Polytetrafluorethylene is made in three steps. In the first step, chloroform is reacted with hydrofluoric acid in the presence of a catalyst, forming monochlorodifluoromethane. This is then pyrolized to form tetrafluorethylene monomer. In the third step, the monomer is catalytically polymerized to form the polymer.

Polychlorotrifluoroethylene. Marketed as Kel-F and Fluorothene, CTFE is a true thermoplastic and can be processed by the usual techniques of injection molding, and extrusion. It has a somewhat lower heat resistance than its sister fluorocarbon, TFE. The ease of molding combined with the stability to thermal and chemical attack has led to industrial applications such as gaskets, chemical tubing, electrical insulators, chemical-tank liners, and electrical insulation.

CTFE is compression molded at 500–600°F and 10,000 psi. It may also be injection molded at 30,000 psi and the same temperature. Corrosion-resistant molds are required owing to slow attack by the molten resin.

The molecule of polychlorotrifluoroethylene resembles that of TFE, differing only in that every fourth fluorine atom in the chain is replaced by a chlorine atom. The resin is made in two steps. The monomer is prepared by dechlorination of 1,1,2-trichloro, 1,2,2-trifluoroethane with zinc dust in boiling alcohol. The technique of polymerizing the monomer, step two, is proprietary but is believed to use peroxides in aqueous suspension.

Phenoxy

Phenoxy is one of our most recent developments in thermoplastics and fills an important niche in providing toughness along with rigidity and good tensile strength. In addition, it can be molded to very close tolerances and has a low thermal coefficient of expansion—about equal to that of epoxy—giving it an excellent combination of properties for good dimensional stability. Its low moisture and gas permeability lend it to container applications, although caution must be used owing to its poor resistance to many solvents. Heat resistance is limited to about 200°F.

Because phenoxy can be blow molded, injection molded, and extruded, it lends itself to fabrication into bottles and similar molded shapes as well as into tubing. The rapid, high-precision injection molding of which phenoxy is capable lends it to items like threaded plastic parts and thin-walled shapes. Molding is usually conducted at 500°F with cylinder pressures in the range of 10,000–15,000 psi.

Phenoxy is manufactured by the reaction of bisphenol-A and epichlorohydrin (also the ingredients for common epoxies). The molecular group is composed of alternating oxygen and carbon atoms and benzene rings, with pendant carbon and hydrogen atoms.

Polyamide

Polyamide, or nylon, is best known for its use as a textile fiber. However, the several molding varieties now manufactured, with their toughness and lubricity, have a variety of applications including chair and drawer glides and tool components. The polyamides are made in a wide variety of compounds, both reinforced and nonreinforced, extending their versatility into electrical, mechanical, consumer, and automotive markets. There are several varieties of polyamide with various properties—"6," "6,6," "610," etc.,—extending the usefulness of the polymer family.

Polyamides may be injection molded or extruded and made into film. Molding cylinder temperatures will range between 350°F, for copolymers, to 530°F for nylon 6,6. A typical cylinder pressure is 20,000 psi.

Polyamide polymers are prepared by reacting a diamine, such as hexamethylene diamine, with a diacid such as adipic. In this instance, the product is nylon 6,6. Polyamides are characterized by having nitrogen atoms in the chain backbone with the carbons, and oxygen atoms interspersed with the hydrogen atoms in the pendant positions. Some varieties have a basic benzene ring with one of the carbon atoms replaced by a nitrogen atom.

Polycarbonate

Polycarbonate is characterized by astonishing toughness. As a result, it is often used as burglar-resistant window glazing, taking advantage of the accompanying transparency. This toughness also has led to usage as small gears, fasteners, and tool components. The most serious weakness of this plastic is its lack of resistance to some solvents.

While polycarbonate has a high heat resistance, it is moldable by the usual thermoplastic processes. Typically, it is injection molded or extruded at 550°F under 16,000 psi barrel pressure.

The molecule of the material is characterized by the presence of a linkage containing one carbon with three oxygens. Manufacture is by a straightforward reaction of phosgene gas with bisphenol-A, the latter being a major raw material used in the manufacture of epoxy resins.

Polyester (Saturated)

The first polyesters developed were of the unsaturated (thermosetting) variety, during the 1930s. These will be discussed in a later section of this chapter. More recently, saturated (thermoplastic) polyesters have appeared and have found broad application as fibers, film, and molded parts and have broken into a major market as blown bottles. Polyester fibers are marketed under such trade names as Dacron or Kodel. The molding compounds are trade named Celanex, Tenite, and Valox. The films are sold under the trade names Mylar and Celanar.

Having good dimensional stability, low moisture absorption, and toughness, thermoplastic polyesters find use in outdoor parts such as sprinklers as well as in electrical parts such as connectors, terminals, and coil bobbins. The injection-molding condition for polyester is usually 460°F at 7,500 psi.

All polyester fibers, films, and molding compounds are similar in nature, being polyethylene terphthalate, a chain having as its basic molecule a benzene ring together with four oxygen atoms. A recent variant of this resin has a very stable aromatic structure and a service temperature of up to 600°F. This is poly p-oxybenzoyl, manufactured by the Carborundum Co. under the trade name Ekonol.

Polyethylene terphthalate is made by reacting ethylene glycol with terphthalic acid or dimethyl terphthalate to form the polymer.

Polyolefins

The polyolefins of greatest commercial importance are polyethylene, introduced about 1940, and polypropylene, first manufactured about 1960. Both of these are soft, waxy, low-melting plastics owing their importance to their comparatively low cost. They are manufactured as fiber, film, and molding compound. Neither plastic has notably high tensile strength, with polypropylene having the higher heat resistance, strength, and stiffness. Both are readily processed by all of the conventional molding processes, and both have excellent resistance to acids, alkalies, and solvents as well as excellent electrical properties. At elevated temperature of 150°F and above, however, both plastics are dissolved by aromatic solvents, and concentrated nitric acid can attack them.

The manufacture of polyolefins is as simple as their molecular structure. Polyethylene is made by polymerizing ethylene gas in the presence of controlled amounts of oxygen at temperatures up to 750°F and pressures up to about 20,000 psi. A modification of this process using catalysts permits substantially reduced temperature and pressure. The catalytic process is also used for the manufacture of polypropylene from propylene gas.

Polyethylene. Because of its excellent chemical resistance and electrical insulating properties, broad use of this plastic is found in chemical ware and as a component of electrical apparatus. Polyethylene is widely used to line chemical drums and as chemical and water piping. Extensively utilized as electrical wire and cable insulation, it is popular as kitchenware for similar reasons.

The most important processing techniques for polyethylene are: coating; rotational molding; thermoforming; extruding; injection molding; and blowmolding. Coatings are normally supplied as powders applied by fusing onto metal surfaces at 500°F. They afford good chemical and electrical resistance for such applications as battery clamps. Rotationally molded parts can be made in very large sizes by allowing powder to fuse onto the inside of heated, rotated molds. A variety of parts the size of a small boat have been manufactured and include signs, bins, mannequins, and other molded objects.

Polyethylene parts made by injection molding, including housewares and chemical fittings, have broad application in areas where keeping cost low is paramount. Injection molding is usually performed at 350°F and with 6,000–10,000 psi barrel pressure. Thermoforming is an important process for the manufacture of shallow containers, plaques, and the like where forms can be stretched out of an extruded sheet under moderate heat and vacuum. Blow molding is very widely used for making inexpensive bottles by extrusion into a hollow mold followed by injection of air pressure that forces the extruded slug into a thin lining conforming to the contour of the mold.

Polyethylene is chemically the simplest plastic, being composed of a simple carbon chain with all of its valence locations being filled with hydrogen atoms. Polypropylene differs from polyethylene in that every other carbon atom has one of its pendant hydrogen atoms replaced by a carbon atom to which are attached three hydrogen atoms.

Polypropylene. This polyolefin can be fabricated by all of the processes used with polyethylene. However, its higher stiffness and considerably higher strength direct it into applications requiring higher durability including such uses as chemical piping and fittings, appliance and household components, and the like. Typical injection molding conditions for polypropylene are 500°F and 15,000 psi barrel pressure.

Polyphenylene Oxide

A product of the technology of the 1960s, PPO is a rigid thermoplastic having outstanding electrical properties. The low electrical-loss characteristic, so important in parts such as radomes and other high-frequency electrical parts, is comparable to that of the soft fluorocarbons such as Teflon and is far superior to that of most other plastics. Its other outstanding properties are high impact strength, very low density (1.06), low creep, and service temperature to over 300°F. Therefore, 36PPO is also suitable for a variety of household and food-processing applications.

Having high temperature resistance, polyphenylene oxide must be molded at a relatively high temperature—on the order of 600°F—and injected with ram pressures on the order of 20,000 psi. It extrudes well in complicated sections and has been processed into film and sheets for electrical applications.

Polyphenylene oxide is manufactured by the oxidation of 2,6-xylenol in the presence of metallic copper. The product is a brownish powder. The molecular structure of polyphenylene oxide is a chain having benzene rings alternating with oxygen atoms. Two carbon atoms pendant from each benzene ring each have three attached hydrogen atoms.

Polystyrene

This well-known thermoplastic is our simplest introduction to "aromatic" compounds. These compounds are named after the first chemicals bearing that name, benzene and naphthalene. As distinguished from the straight-chain "linear" chemicals comprised of carbon atoms linked to one another in a chain and having little distinguishing odor, the aromatics have a strong distinguishing fragrance of which naphthalene (moth balls) is a good example. The aromatics have six carbons arranged into a ring formation, the well-known benzene ring. To these carbons can be attached other carbons, or hydrogens, or oxygens, etc. In turn, the rings can be joined together into chains. Usually, such chains yield plastics with improved properties over the linear plastics.

In its pure form, polystyrene is a water-clear, rather brittle plastic extensively used for containers owing to its clarity. Because of its good electrical properties, it is occasionally used as electrical insulators, although its brittleness limits this application. An increasingly broad use of polystyrene is as a blend with acrylonitrile and butadiene to form ABS, a versatile, tough plastic formulation widely found today in the appliance and automotive markets.

Polystyrene is a relatively hard plastic with a moderately low melting point, about 350°F, and is readily attacked by solvents. It is resistant to acids, alkalies, alcohols, and most oils.

The molecule of polystyrene is comprised of benzene rings alternating in a chain with single carbon atoms each having two attached hydrogen atoms. Polystyrene is manufactured by the polymerization of styrene monomer under moderate heat and pressure, using peroxide as an initiator. The three process variables permit a wide variation of molecular weight among polystyrene products.

Polysulphone

Polysulphone is the first of the sulfur-containing plastics, and the molecule superficially resembles that of polystyrene in that it is comprised of a chain of benzene rings. However, instead of having alternating carbon atoms, as in the case of polystyrene, the basic polysulphone molecular chain is comprised of a benzene ring followed by a carbon atom, another benzene ring, an oxygen atom, a ring, a sulfur atom, a ring, and an oxygen atom. This basic molecule is repeated a number of times to form the plastic. Polysulphone was developed by Union Carbide Co. at its Boundbrook Laboratories after World War II and was introduced commercially in 1965.

Owing to its excellent heat resistance, polysulphone sees applications such as "under-the-hood" automotive parts, electrical parts subjected to soldering, and hot appliance parts. It is normally injection molded with a barrel temperature in the 625–750°F range with pressures of about 15,000 psi.

The complexity of the polysulphone molecule results in an unusual plastic often used in applications normally requiring thermosetting plastics because of demands of temperature or dimensional stability. Polysulphone has a melt temperature of 600°F and a service temperature of about 325°F and is highly chemical and solvent resistant. Owing to its very high melt temperature, the plastic is relatively complex to mold, and its applications, which are cost limited, have tended to be confined to technical and industrial areas. The manufacturing process is proprietary to its producer, the Union Carbide Co.

Polyurethane

The polyurethanes are actually elastomers and are made in two forms, the thermoplastic and the thermosetting. In both forms, they are characterized by toughness and abrasion resistance.

Thermoplastic urethane may be processed by injection molding and extrusion and has moderate heat and solvent resistance. Its unusually high abrasion resistance, softness, and high elongation have pointed its applications toward wear-resistant products such as shoe heels, spouts, and similar uses where conventional plastics would not last very long.

Thermoplastic urethanes are normally injection molded at 375–425°F and at only 400–1,000 psi pressure. The thermosetting compression-molded formulations, as distinguished from the casting rubbers, are cured at 345°F and moderate pressure followed by an hour-long postcure at 482°F.

In the thermosetting form, polyurethanes have found their major applications as both flexible and rigid foams—perhaps the most significant aspect of urethane foam is the ease with which it can be made into a wide range of densities and hardnesses. It therefore lends itself to applications such as cushioning materials as as well as to thermal insulation and structural blocks.

Reflecting their versatility, polyurethanes can be continuously machine formulated or batch mixed. In commercial manufacture of polyurethane foam, the ingredients are proportioned by a mechanical device and continuously blended and pumped into a moving trough. As the trough moves along, the polyurethane foams up, filling it. The foam reaches the end of the machine as a cured loaf, ready for slicing by band saw into sheets and blocks.

For void filling, the polyurethane raw materials may be batch mixed, usually with a propeller mixer, and poured into a particular cavity before the batch has time to rise. Once in the cavity, the material rises and entirely fills the space. In this manner, products such as boat hulls are successfully made from rigid foam.

Polyurethanes are manufactured by reacting a polyfunctional alcohol with an isocyanate in a simple blend-reaction operation. By the proper choice of the alcohol and the isocyanate, polyester or polyether urethanes can be made. Similarly, either thermoplastic or thermosetting urethanes can be made by the proper choice of isocyanate.

Vinyl Resins

The vinyl resins are characterized by a basic molecule having two carbon atoms with three pendant hydrogens and a fourth pendant group that is a unique group of atoms depending on the particular vinyl. It is this fourth group of atoms that provides the special qualities of the particular vinyl resin. The basic molecule repeats in a chain to form the plastic polymer.

Polyvinyl Acetate. Polyvinyl acetate first saw commercial manufacture about the end of World War I but was never widely produced as a molding material owing to its low softening temperature, about 200°F. However, because of its good adhesive properties and toughness, it has seen use in the manufacture of floor tile, pressed-wood-particle board, and artificial leather. Today, its main use is probably as a general-purpose adhesive marketed as white glue and hot-melt adhesive.

When acetylene is reacted with acetic acid in the presence of a catalyst at about 100°F, vinyl acetate is formed. This is a colorless liquid with a boiling point above room temperature. The polymer is then formed in a solvent under heat and pressure, precipitated, and dried.

Polyvinyl Alcohol. Polyvinyl alcohol is produced commercially as a white powder soluble in water or alcohols. In its cast form, PVA is characterized by water solubility, toughness, low gas permeability, high resistance to organic solvents other than alcohol, and crystallinity when stretch oriented. Crystallinity allows the material to polarize light. A series of hydrolysis levels of the resin are available, allowing the user to select materials of room-temperature solubility ranging from highly soluble to not soluble at all.

Conventionally, PVA is cast into sheet form from solution and plasticized with glycerine to confer toughness. Its products are characterized by high tensile strength and good tear and abrasion resistance. The polarizing characteristic of the stretched material has led to its use in sunglasses trade named Polaroid.

In a typical manufacturing process, polyvinyl alcohol is swelled in anhydrous alcohol. The reaction is then produced under a moderately elevated temperature in the presence of a low level of either an acid or an alkali. The process may be interrupted at a given point to produce a product of a predetermined hydrolysis (hence solubility) level. There exist a variety of these levels on the commercial market.

Polyvinyl Butyral. This resin is characterized by its clarity, toughness, and adhesive properties. It has a tensile strength of about 8,000 psi, but its heat distortion is only about 125°F, which limits its applications. It has no "snap back" but is dead, with damping qualities reminiscent of silicone rubbers.

By far the largest use of polyvinyl butyral is as the inner ply of safety glass, which application it has dominated since World War II because of its unique qualities. The other major market for butyral is as an additive to adhesive that contributes great toughness and elongation. In this application, it has found major usage in structural adhesives for aircraft. The manufacturing process is similar to that for the manufacture of polyvinyl acetate.

Polyvinyl Chloride. Polyvinyl chloride is commercially the most important of the vinyls because of its many uses. Indeed, the word "vinyl" is most commonly applied only to the vinyl chloride formulations. Polyvinyl chloride was first polymerized in 1872, but it was not commercially manufactured until 1936 and was not widely produced until the technique of plasticization was developed to yield stable, tough, flame-resistant, rubberlike materials.

In its unplasticized form, polyvinyl chloride is a hard, horny resin lacking essential toughness. Owing to its chemical resistance, it sees use as sheet rod and tube for use in plating, chemical manufacturing, and chemical transport.

Plasticized PVC is prepared by blending the resin with oils, such as tricresyl phosphate and dioctyl phthalate, or with polymeric plasticizers which confer superior stability but are more expensive, such as sebacic acid polyester. The plasticized form of PVC is flexible but lacks the resiliency of rubbers. It begins to soften at about 150°F.

Both the plasticized and unplasticized forms of the resin may be injection molded or extruded. In unplasticized form, it is commonly heat welded to form chemical tanks. In plasticized form it can be blown into film and coated from a solventless fluid known as a *plastisol* to form gloves and similar objects. Injection molding of unplasticized PVC is conducted with barrel temperatures in the range of 280–340°F and a pressure of 12,000 psi.

Vinyl plastisol is a dispersion of the vinyl resin powder in a plasticizer with appropriate fillers and pigments. Plastisol is a stable liquid that can be formulated in a range of viscosities. In liquid form, it can be poured into molds, coated on fabrics, sprayed, or applied by dipping. Heated to about 300°F, the plasticizer dissolves the vinyl powder, and the solution, when cooled to room temperature, then forms the flexible vinyl known to commerce. Typical applications are upholstery, sink racks for dishes, film, protective gloves, shoe soles, floor tile, garden hose, and gaskets.

In the typical manufacturing process, acetylene is passed through a hydrochloric acid bath in the presence of a catalyst to form vinyl chloride. To form the polymer, the monomer is heated to about 125°F in the presence of free-radical-forming compounds such as azos or aryl peroxides. The polymer precipitates as a powder and is then washed and dried.

THERMOSETTING RESINS

Thermosetting resins have come a long way since the dark-colored, water-producing phenolics of Dr. Baekeland's day. Now, water-white, color-stable, low-shrinkage, and highly dimensionally stable thermosetting resins are available. Over the years, we have learned to mold the thermosets by compression, transfer, and even extrusion so that, except for having a cure cycle, they are almost as versatile as the thermoplastic resins. We can choose them in prices ranging from low-cost phenolics to expensive polyimides in a range of colors and properties.

Alkyd

Among the oldest molding compounds, the alkyds have been manufactured since World War II. Their primary advantages over formaldehyde-based resins are dimensional stability, superior electrical qualities, and the fact that they do not yield (condense) water during cure. As a consequence, they have found a ready market in the electrical industry, including automotive applications. Alkyd resins are actually an offshoot of the alkyd-coating industry introduced to the American market from Germany. Of moderately high heat resistance, rapid cure cycle, and good mold flow, alkyds permitted relatively complicated molding and were rapidly accepted.

The term *alkyd* should be used with some care. It was originally coined to designate resinous reaction products of di- and polyhydric alcohols and acids, which, when modified with reactive monomers, are termed *polyester resins*. It is when they are modified by formulating with nonvolatile monomers such as diallyl phthalate to form dry, or relatively dry, molding compounds that they are termed "alkyd plastics." It will be noted that the alkyd plastics are actually polyesters. Indeed, in the trade, the basic polyester, uncut with a monomer such as styrene, is called an alkyd.

The alkyd molecule is best represented by alternating alcohol and acid molecules having pendant oxygen atoms that permit reaction with fragments of peroxide during cure for cross-linking with monomers.

Alkyd molding compounds are formulated with a variety of fillers and reinforcements, including clays and carbonates, fiberglass and Dacron. They may be compression or transfer molded at pressures to 10,000 psi and at temperatures that depend on the peroxide and monomer used but they generally range from 200–350°F.

Alkyd resins may be manufactured from virtually any di- or polyhydric alcohol reacted with an acid. A typical alkyd is made from phthalic anhydride and ethylene glycol under ambient pressure and elevated temperature. To prevent the reaction from going out of control, it is necessary to prevent the presence of oxygen, and normally an inert gas such as carbon dioxide is bubbled through the reactor to sweep out the moisture products of reaction as well as oxygen. After the alkyd is formed, it is blended with a monomer and an inhibiter, and then combined with fillers, pigments, and reinforcements and granulated.

Allylic

A post-World War II development, the allylic resins are among the more versatile of the thermosetting resins and possibly the least appreciated. Actually a branch of the polyester family, the allylics can be homopolymerized or copolymerized with alkyds. Easily molded, they can be cured at any temperature above 200°F as homopolymers by the proper selection of peroxide, or at temperatures considerably lower by cross-linking with alkyd. Using ultraviolet-sensitive accelerators, they may be sunlight cured as alkyd copolymers.

Molding processes include contact, vacuum and pressure bag, compression, and transfer molding.

The earliest allylics were formulated in the post-World War II period but found little application owing to the extreme difficulty of curing the liquid monomer. The monomer may be homopolymerized in the presence of a peroxide, but without almost exhaustive processing cannot be fully cured. It was eventually discovered that it was possible to cure the monomer partially and then precipitate it to form a powder of molecular weight intermediate between a monomer and a fully cured polymer. This powder was termed a prepolymer and led to one of the earliest stable pre-preg materials in the late 1940s.

As a result of the prepolymer development, allylics are marketed as either monomer or prepolymer. The processer is then free to formulate compounds with any ratio from zero to over 50% monomer, blended with appropriate peroxide, reinforcement, filler, pigment, and copolymer.

The only example of a truly cast allylic homopolymer cast from the monomer is diethylene glycol bis (allyl carbonate), which is cast and sold under the trade name CR-39 for optical use as sunglasses. It is becoming increasingly popular for conventional eyeglasses because of the unusually large lenses that are now the craze. CR-39 is characterized by good optical transparency and excellent hardness and scratch resistance.

Diallyl phthalate is most widely used in combination with alkyd resins for applications such as "polyester" pre-preg. The nonvolatile, stable diallyl phthalate monomer confers good shelf life, excellent laminating qualities, and good surface hardness. A common example of its use is the imitation-wood-grain low-pressure laminate used for low-cost furniture and wall paneling. Such pre-preg contains 5 to 10% DAP. Typical fiberglass "polyester" pre-preg used for aircraft and similar structures also contains substantial levels of DAP monomer.

Diallyl orthophthalate, diallyl metaphthalate monomers, and prepolymers are widely used in precision compression moldings for such applications as electrical connectors and radomes. They are chosen for these applications for their outstanding dimensional stability (hardly excelled by any other thermoset), low water absorption, outstanding electrical properties, and ease of molding. Diallyl orthophthalate (diallyl phthalate) is used for service at temperatures up to 300°F. For superior dimensional and thermal stability, diallyl metaphthalate (diallyl isophthalate) is usually selected. The two isomers have equivalent ease of molding.

Diallyl phthalate pre-pregs are laminated at 200–400°F at pressures from vacuum bag to 150 psi. Molding compounds are usually processed at 300°F and 500–10,000 psi.

The most important allylic molecule, that of diallyl phthalate, is a chain of six-sided benzene rings. From two carbons of the ring are pendant allyl groups—four carbon chains with an intermediate oxygen atom. There are several other allylic resins, which, like CR-39, have the two allyl groups joined by molecules other than benzene rings.

Allylics are manufactured by a condensation reaction of an alcohol and an acid. Diallyl phthalate, for example, is reacted in the presence of an inert gas from phthalic anhydride and allyl alcohol to form the orthophthalate. The resultant monomer has a high boiling point and is usually stripped of low-boiling contaminants under vacuum.

Prepolymer is made from the monomer using a mixed peroxide system in a reaction vessel. The reaction terminates at 25% prepolymer yield, and then it is precipitated using alcohol, and washed and dried.

Epoxy

Epoxy resins are surely the most versatile thermosetting resins in our inventory. They are strong, have outstanding properties, may be formulated as liquid or solid resins, and are available in a range of states from flexible to rigid and with temperature resistance to about 400°F.

The two largest applications of epoxies are adhesives and laminating resins, and the third largest is coatings. Castings form the smallest application for these versatile resins.

Epoxies first commercially available in Germany in the 1930s. Following World War II, they appeared in Switzerland and the United States, and in the 1950s they began to be used as coatings and casting resins despite their high price. Today, epoxies are widely found in structural adhesives, structural composites, and fiberglass as well as in more exotic materials like graphite fibers and boron. Printed circuit boards using epoxy monopolize the marketplace. Since the epoxies are available in a range of viscosities from solid to watery and can be cured with a variety of agents, it is common for the end user to formulate a system to suit specific requirements. Hence, there are dozens of organizations that do nothing but formulate epoxy systems for resale as casting or laminating resins under individual trade names. There are also an additional dozen or more formulators who make molding compounds or preimpregnated fabrics and tapes. Yet only two or three prime manufacturers of epoxy resins exist, and government restrictions threaten to reduce their number further.

Epoxies may be cured at room temperature using amine catalysts or, for superior thermal and physical properties, cured at up to 350°F using amines or anhydrides. Laminates are commonly cured at 300°F with pressures ranging from vacuum bag to 300 psi. Molding compounds are typically cured at 300°F and 500–2,000 psi.

The most important epoxies are the epichlorohydrin–bisphenol-A types. Bisphenol-A is first prepared by reacting phenol and acetone. Epichlorohydrin is produced directly from the petroleum fraction propylene. In the presence of a catalyst, bisphenol-A and epichlorohydrin react to form a long chain molecule having carbon atoms and benzene rings interspaced along the backbone. In each molecular unit there are two pairs of carbon atoms, each pair of which shares an oxygen atom to form the epoxide group. It is this group that opens in the presence of a curing agent to form the cross-linked, cured epoxy resin.

Novolac epoxies are higher-temperature resins characterized by benzene rings alternating with ethyl (CH_2) groups. Each benzene ring has a pendant group that includes an epoxide group. Accordingly, the novolac system has more epoxy groups per molecule than the "epi-bis" epoxies and thus exhibits more cross-linking during cure as well as higher heat resistance.

Melamine

The melamine formaldehyde resins, first commercialized just before World War II, are characterized by outstanding hardness, stain resistance, density, and color stability. As a result, they have found major applications as houseware, furniture, and wall paneling. For many years, dishware has been made from filled melamine which features all the melamine properties, particularly scratch and stain resistance. Simulated-wood tabletops and wallboard, popularly known as Formica, is often actually melamine. Other applications include buttons, housings for small appliances, and electrical switchgear parts. For a long time, electrical connectors were satisfactorily made from melamine, but with the advent of ultrasmall connectors requiring very good dimensional stability and accuracy, the allylic resins gradually replaced the melamines.

Melamines are readily molded at temperatures in excess of 300°F and pressures ranging from 2,000 to 20,000 psi. Owing to the evolution of gases during cure, the mold must be momentarily opened or "bumped" during the cycle. Both compression and transfer molding methods are used.

The melamine formaldehyde molecule is essentially composed of six-carbon benzene rings with alternating carbon atoms on the ring replaced with nitrogen. To the three remaining carbon atoms are attached molecules comprised of nitrogen, hydrogen, and carbon.

Melamine formaldehyde resins are made in a multistage operation starting with calcium carbide, which is reacted with nitrogen to form calcium cyanamide. Reacting the calcium cyanamide with carbon dioxide, dicyandiamide (dicy) is formed. This substance is further polymerized under heat and pressure to form melamine. If melamine is next reacted with formaldehyde, the monomer of melamine formaldehyde, methylol melamine is formed and this can be converted into melamine formaldehyde under mild heating.

Phenolics

The phenolics include an extraordinary variety of resin systems all commonly referred to as phenolics, although they may be considerably different from one another. Among the phenol formaldehyde resins are many molecular configurations and properties, so that it is unwise to assume that all phenolics have the same properties.

Based on work done many years earlier by Dr. Baeyer, phenolics were developed in about 1909 by Dr. Baekeland. By about 1922, the phenolic resins were established as plastics, although not without difficulty, and by the end of the 1920s there were a number of manufacturers in the field. Very versatile materials, phenolics are used as coatings, casting resins, foams, laminating resins, and compression molding resins. When properly formulated, phenolics have unusually high heat resistance. Indeed, a 500°F fiberglass-reinforced phenolic introduced in the 1950s and sold at a very high price was actually based on an inexpensive casting phenolic. Among the disadvantages of the phenolics are relatively poor water resistance and poor color stability.

The applications of phenolics are, as might be expected, quite broad. As laminating resins they are found in aircraft structural parts and in aircraft turbines when reinforced with fiberglass or graphite fibers. They have been used as simulated-wood-grain wallboard where dark colors and lack of light stability were not important.

As molding compounds, phenolics are found in all kinds of household appliances, although they are being rapidly replaced by thermoplastics. Phenolics are also commonly used as distributor caps and in other engine parts in automobiles as well as in numerous industrial applications such as pulleys, knobs, switch bases, and terminal blocks. As casting resins, they have been used for tooling patterns for many years.

Phenolics may be the most important commercial binding resins and, indeed, this may be their largest current area of use. Foundry core binder is a major application in which the liquid or powder resin is used to bind sand together into shapes that will eventually become the hollows in metal castings. Starting in the late 1950s, the lumber industry found that instead of burning up the wood chips resulting from lumber production, they could granulate them and bind them together with phenolics to form particle board. In the last 20 years, the market for this product has grown astronomically and can now be measured in the billions of sq ft per year.

Readily molded under heat and pressure, phenolics are typically laminated or compression molded at 270–400°F with molding pressures up to 5,000 psi.

The typical phenolic molecule is a chain of benzene rings alternating with carbon atoms. To one carbon of each benzene ring is attached an oxygen atom with a pendant hydrogen.

There are far too many methods for the manufacture of phenolics for complete discussion here. Basically, however, phenol is reacted with formaldehyde to form phenol formaldehyde. Depending on the catalyst and the proportions of the basic ingredients a very wide range of molecular configurations and properties result.

Polyesters

More accurately termed *unsaturated polyesters* because they react with monomers, the polyesters represent the largest segment of the reinforced plastics market, with applications ranging from the highly sophisticated to the mundane. As complex as the phenolics in their possible variations, polyesters can be made with toughness on one end of the spectrum and brittleness and heat resistance on the other. They may be formulated with a range of electrical properties and chemical resistances, and formulations having excellent sunlight resistance are available.

A popular, high-production application of polyester using a tunnel-oven process is translucent sheeting reinforced with fiberglass. Awnings and fencing are typical applications. Another very large application of chopped-glass-reinforced polyester is automobile bodies; for example, the Chevrolet Corvette. This type of polyester is also used in the manufacture of aircraft radomes, chemical equipment, appliance housings, and electrical insulation. Polyester molding-compounds called premixes are widely used for appliance parts and electrical motor insulators.

Polyesters are commercially prepared with a monomer as blends of the basic polyester, usually termed an "alkyd." This blend is stabilized with a chemical inhibiter to prevent premature hardening. Owing to cost, the monomer most popularly used is styrene. However, to produce other improved properties, other monomers can be used; for example, methyl methacrylate for better color, vinyl toluene for lowered volatility, diallyl phthalate for improved electrical and chemical properties, or triallyl cyanurate for increased resistance. The monomer level is usually in the range of 30 to 50%.

Curing is effected by the decomposition of an added peroxide, usually by heat. The peroxide most commonly used for heat cure is benzoyl peroxide. However, room temperature curing may be accomplished by decomposing the peroxide chemically through a "redox" reaction. The typical agent for this purpose is cobalt naphthenate used in combination with methyl ethyl ketone peroxide (MEKP).

In casting, the low-reactivity polyesters must be used to avoid a too-rapid chemical reaction. The combination of a low-reactivity alkyd with a light, stable monomer and the proper redox agent can yield an almost water-white casting of the type popularly used for hobby crafts dealing with the making of paperweights and other decorative items.

Alkyds, which are members of the polyester family, are chemically characterized by the ester linkage in the resin chain. This linkage is a double bond between one of the carbons and an oxygen atom. In the curing process, the peroxide catalyst is broken down by heat or a redox process. Fragments of the peroxide then displace the oxygen atoms and join elements of the alkyd and monomer to form the final thermosetting structure.

Polyester alkyds are made by the reaction of a polybasic acid and a polyhydric alcohol. The reaction is conducted in the presence of heat but without pressure. While the possible combinations of acids and alcohols that may be used are endless, a typical general-purpose (medium-reactivity) polyester will be reacted using a 1:1 blend of phthalic and maleic anhydrides, together with propylene glycol. If a more heat-resistant, harder polyester is desired, the phthalic anhydride may be reduced in proportion or omitted to yield a high-reactivity resin. Conversely, elimination of the maleic anhydride will produce a tougher, castable, low-reactivity resin. If more rubberlike properties are desired, sebacic or adipic acids may replace the anhydrides, or isophthalic acid may be used to obtain an intermediate toughness.

After the alkyd reaction is completed, the viscous alkyd is pumped into a blending vessel containing the monomer, the blend is prepared, and an inhibiter is added to prevent premature jelling. In the inhibited form, the shelf life of uncatalyzed polyester is one or two years.

Polyimide

The polyimide resins are quite recent polymers that are known for their heat resistance and stability. In their various forms, they have heat resistance up to 900° F for the short term and up to 600° F continuously.

A number of polyimides have been introduced, remained in the marketplace for a short time, then disappeared owing to processing difficulties or lot-to-lot variation in properties. Polyimides are typically difficult to process, requiring curing temperatures of up to 550°F and pressures ranging from 100 psi to several thousand psi. Most also require a severe postcure to develop full properties.

Polyimides have appeared on the market as unfilled blocks and rods, as film, and as laminating and molding resins. The block form finds use as high-temperature electrical and thermal insulation as does the film variety. The laminating resin has found its major applications in printed circuit boards and in high-temperature structural laminates.

Typical polyimides are made by the reaction of a dianhydride with a primary amine. As an example, pyromellitic dianhydride can be reacted with p-phenylene diamine to form a thermosetting amide. The polyamide molecule is characterized by the presence of the amide group, a five-side ring with carbon in all of the positions except one which is occupied by a nitrogen atom. From the two adjacent carbon atoms, there are pendant oxygen atoms.

Silicone

Silicone resins were first manufactured during World War II by several American sources and have found substantial use ever since in areas where retention of properties at both high and low temperatures is required. The temperature range −70° to 600°F is generally covered with little loss of properties.

Silicones are manufactured in a variety of types, including oils, greases, coating, rubbers, molding compounds, casting compounds, and laminating resins.

Silicone rubbers are unusual because they are vulcanizable at either room or elevated temperature. Owing to the ease of curing and handling, the room-temperature-vulcanizable (RTV) rubbers have attained wide popularity and are used in potting of electrical components, for plastics molds, and other potting applications. The tooling application is particularly interesting owing to the high fidelity of the dimensions, the heat resistance, and the remarkably high thermal coefficient of expansion that can be used to advantage to provide molding pressures. The elevated-temperature-vulcanized rubbers are commonly used for gaskets, seals, ducting, and electrical insulation because of good electrical properties and resistance to chemicals and oils. These rubbers are molded at about 300°F at moderate pressure in compression tooling.

Silicone laminating resins are used with fiberglass reinforcement when retention of physical properties up to 600°F is important. Possibly the most frequent application of silicone laminating resins is in electrical insulation, but they are also used in hot-air ducting in aircraft.

Although they have been used for electrical connectors and terminal posts, silicone molding compounds are not particularly widely employed. Laminates can be made at vacuum-bag or higher pressures and are generally cured at temperatures in excess of 300°F, depending on the particular resin.

The basic silicone chain is comprised of alternating atoms of silicon and oxygen. Pendant from the silicon atoms may be a variety of groups, depending on the particular type of silicone. The most important of these types is dimethyl silicone, where two carbon atoms are attached to each silicon atom in the chain, and three hydrogen atoms are attached to each carbon atom.

The simplest process for the manufacture of dimethyl silicone is by reacting methyl chloride with silicone in the presence of copper. Several refining and reaction steps lead to dimethyl silicone, which is then formulated into appropriate compounds.

Urea

Commercial production of urea formaldehyde dates from the early 1930s when molding compounds, laminating resins, and coatings were developed. At one time, urea adhesives were popular for use with plywood, but more highly water-resistant phenolics have largely replaced them. Owing to their light color, particularly when reinforced with alpha cellulose, the urea resins were once very popular for appliance housings. Highly translucent, pearllike qualities in a variety of attractive colors are quite possible with ureas. However, as the tougher thermoplastics with rapid molding cycles were developed, urea resins were largely displaced from their early markets. Buttons remain their major foothold.

Urea resins are compression or transfer molded at 275–340°F and 1,500–2,000 psi. Because water is emitted during cure, it is generally necessary to open briefly or to "bump" the mold during the early part of the cycle, in order to release this vapor.

Even though urea formaldehyde resins have been manufactured for almost 40 years, their chemistry is not completely understood, and many variations of the basic molecule can result by slight changes in the manufacturing process. Basically, urea is reacted with formaldehyde in the presence of a catalyst. The resultant resin is then filtered out and spray dried, after which it is blended with appropriate fillers and packaged. The molecule of urea has a backbone of carbon and nitrogen atoms, while pendant from the backbone are carbon, oxygen, nitrogen, and hydrogen atoms.

3 Injection Molding, Blow Molding, Extrusion

INJECTION MOLDING

INJECTION MOLDING MACHINES

The vast growth of injection molding is reflected dramatically in the many types and sizes of equipment available today, and a complete catalog of all available makes and models is beyond the scope of this chapter. The wide choice of equipment options is illustrated by the following summary of principal types

Injection Unit (See Figures 3-1, 3-2, and 3-3.)
1. Conventional (one-stage) single-plunger
   a. Horizontal cylinder
   b. Vertical cylinder
2. Reciprocating screw—plunger
   a. Hydraulic drive for screw rotation
   b. Electric drive for screw rotation
   c. Screw preplasticizer, hydraulic drive; plunger injection

Mold Clamping Unit
1. Hydraulic cylinder, horizontal clamp
2. Toggle clamp
   a. Horizontal
   b. Vertical

In addition to the principal combinations indicated above, injection equipment manufacturers can provide various additional features, such as:

1. Rotating spreader
2. Ram speed control by flow control valve or by booster hydraulic pump

3. Two-stage injection pressure, individually controlled
4. Adjustable, two-speed mold opening/closing stroke
5. Internally heated torpedo
6. Plunger prepositioning
7. Multiple prepack or "stuffing" stroke control
8. Proportioning-type cylinder zone temperature controllers
9. Variable speed adjustment of plasticizing screw
10. Various screw designs, including valved screw

Selection of the specific type of injection equipment is governed largely by the requirements of the molding job and by the degree of versatility desired. Each type of injection unit and clamping unit has its advantages and limitations and, therefore, its advocates and critics. The current trend is toward reciprocating screw and two-stage screw preplasticizer equipment, primarily because of its increased capacity and versatility compared with conventional plunger machines, even though the initial cost of screw equipment is usually 20 to 30% higher.

The major advantage of a toggle clamp unit over a hydraulic clamp unit is that it permits faster machine cycling. However, toggle clamping is generally limited to 400 tons maximum clamping force and thus finds its principal use in the smaller, fully automatic molding operations.

Regardless of type, injection molding machines with the following

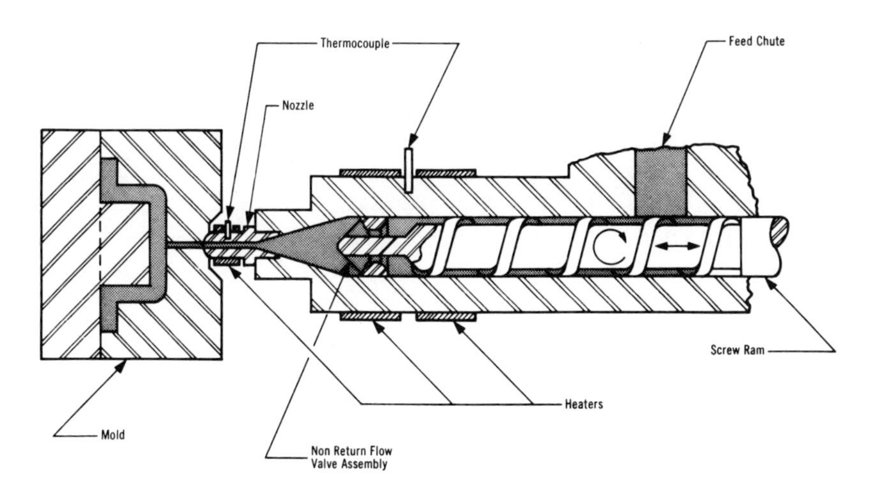

Figure 3-1. Screw ram cylinder section.

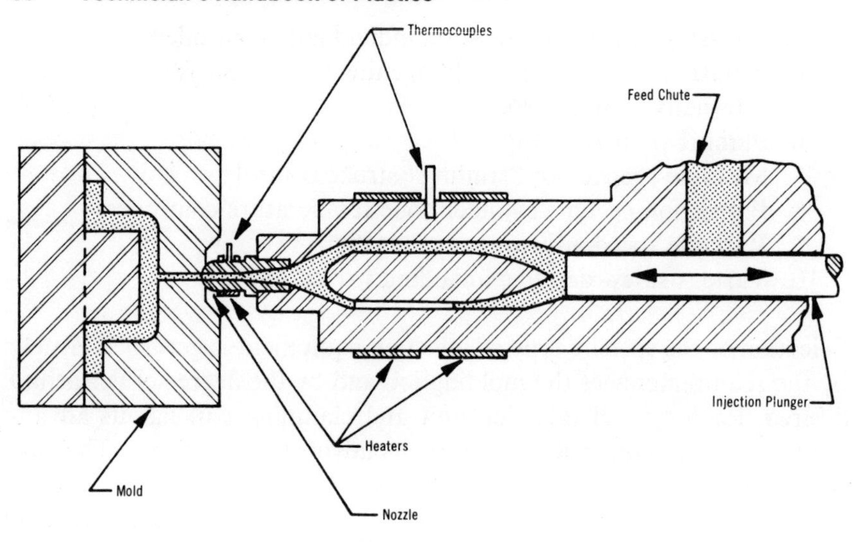

Figure 3-2. Plunger-type cylinder.

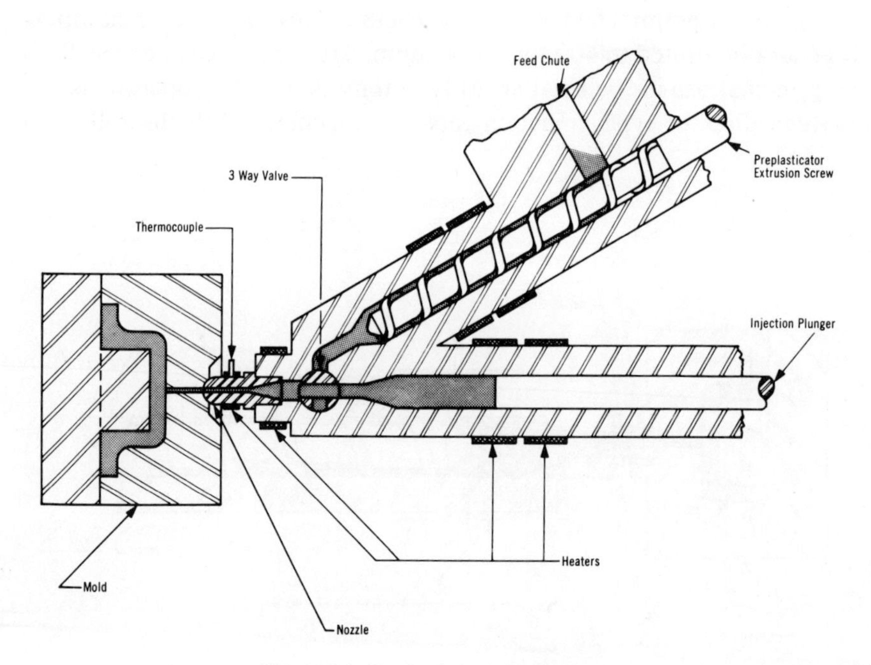

Figure 3-3. Preplasticator unit section.

features are desirable for obtaining best equality and versatility of operation:

1. Injection molding cylinder: At least three individually controlled heater zones, preferably with proportioning type or dual-wattage temperature controllers. Nozzle temperature should be separately controlled, either by individual pyrometer or by rheostat.
2. Injection pressure: Variable up to at least 20,000 psi exerted by the plunger face, preferably with two stages (injection full pressure and injection holding pressure), each controlled by individual timers. Pressure-actuated solenoids are available for maximum efficiency.
3. Variable ram speed, either by flow control valve or by timed booster hydraulic pump, with maximum speed of at least 60 in./min.
4. Precision feed adjustment, accurate to permit shot weight control within 0.01 oz.
5. Cycle element timers, accurate to within 0.1 sec, to control plunger forward time, dies closed time and/or total clamp time.
6. On conventional plunger machines, plunger preposition and multiple prepack control.

Preheating and Predrying

Predrying provides the additional benefit of preheating the molding material, thereby reducing the amount of heat input demand on the injection cylinder heating bands. Without preheating, the cylinder must heat the pellets from about 70°F to, say, 470°F—a differential of 400°F. By preheating to 190°F, this differential is only 280°F, and the heating demand on the cylinder is therefore reduced 30%. This provides more uniform melt temperature, reduces the possibility of localized overheating (i.e., "hot spots") in the cylinder, and frequently permits faster molding cycles. Preheating is especially advantageous in those cases where the molded shot weight exceeds 70% of the rated machine capacity. For these reasons, preheating is highly recommended.

Both oven/tray dryers and hopper dryers are suitable. Hopper dryers offer the distinct advantages of reduced handling and less possibility of contamination, and keep the material hot in the machine hoppers. The use of dehumidified air is preferable, particularly when high-humidity conditions exist.

Pellet Geometry

Molding polymers are usually fed to the molding machine in either pelleted or granular form. Pellets are small regular shapes, generally cubical or cylindrical, and most often produced by cutting extruded strands to consistent lengths. Granules are irregularly shaped particles produced by crushing polymer chunks or slabs, or by grinding sprues and runners; they come in a wide range of particle sizes.

When molding machines are fed by volumetric displacement feed mechanisms, a decrease in bulk density requires a greater feed setting. This is because more air must be pressed out in compressing the plastic feed in the cylinder. With materials having a fairly low bulk density, the plunger may even have to tamp in several strokes to accumulate sufficient plastic to fill the mold, a procedure called *prepacking* or *stuffing*. If the trapped air does not escape as the material densifies in front of the plunger, it may become trapped in the molten polymer and be carried into the mold. Here it will appear as splay marks, silver streaks, mica flecks, bubbles, blisters, or black streaks.

Fine pellets have much more surface area per lb than regular pellets. The amount of surface area per lb of pellets is important in dry-coloring and also in use of external lubricants. Fine pellets, having a higher ratio of surface area to weight, provide more uniform distribution of pulverized or liquid additives.

In actual practice, the use of very fine particles is limited by their tendency to overheat and scorch in the injection cylinder. For example, plastic dust from scrap granulators can become trapped in the clearance between the plunger and the cylinder on conventional plunger machines, where it is subjected to scorching by the rear-zone heaters when the plunger is in its forward position. For styrene-based polymers, the minimum practical pellet size for plunger machines is about 0.040 in. $\times$ 0.040 in.

When molding granular fines or fine pellets, it is preferable to maintain the rear zone temperature setting about 30–50°F lower than the center and front zone temperature settings. This procedure not only reduces the tendency to scorch, but also allows the trapped air to escape more readily.

Lubrication

External lubricants are applied to the surface of pellets and granules to allow them to flow more readily in machine hoppers and feed chutes. Also, in conventional plunger machines, a pressure loss occurs in compacting and propelling the hard unfused pellets in the rear of the cylinder. This pressure drop may be as high as 50% of the pressure applied by the plunger face. External lubrication on the pellets or granules makes them more slippery, thus reducing this pressure drop significantly. Surface lubricant also minimizes black streaking caused by frictional scorching.

During the molding process, the original external lubricant is compounded into the plastic material, such that scrap regrind no longer possesses any external surface lubricant. For this reason, granulated scrap may show black streaks or exhibit stiffer flow than its parent virgin compound. In such cases, the molder may need to add fresh external lubricant to his regrind before feeding it to the molding machine. This can be done easily by adding 0.03% powdered lubricant to the regrind and drum-tumbling for 15 min. This percentage is equivalent to 13 grams lubricant per 100 lb of regrind. Excessive external lubricant, or nonuniform blending may cause lubricant streaks or smears on the molded articles.

Mold Surface Temperature

High mold surface temperatures reduce orientation stresses, provide higher surface gloss, and minimize weld lines and flow marks. Unfortunately, these high mold temperatures also require longer molding cycles in order to "set up" the plastic to a sufficiently rigid state to be ejected. Low mold temperatures, which permit faster cycles, have the serious disadvantage of causing high molded-in stresses, poorer gloss, and more prominent weld lines. Cold molds are also prone to cause cracking on ejection when molding the less ductile styrene-based materials.

The *maximum* mold temperature for any amorphous thermoplastic is about equal to its ASTM Heat Distortion Temperature (ASTM D-638). Above this temperature, the molded article will not retain its desired shape and dimensions on ejection. The *minimum* mold surface temperature suggested is about 70–80°F; lower temperatures cause high molded-in strains (from both orientation and thermal shock), warpage, and dull surface finish. *Optimum* mold temperature is a compromise between these two extremes.

Note that these are mold *surface* temperatures, not the temperature of the water circulating through the channels in the mold. On fast cycling jobs, it is sometimes necessary to circulate refrigerated water through the mold in order to remove heat rapidly. In such cases, however, the temperature of the *cavity surface* should not be under 70–80°F.

An effective method for attaining shorter cycles without sacrificing gloss is to maintain the mold halves at different temperatures. The mold half that forms the appearance surface of the molded article (usually the cavity half) can be held at a temperature 20–60°F, higher than the mold half forming the nonappearance surface (usually the forces, or ejection, half). The cooler core promotes rapid chilling and set-up, while the hotter cavity provides high gloss. This procedure should be tried cautiously, for two reasons:

1. Thin flat areas are prone to warp after ejection, so that the "hotter" surface will become concave.
2. The hotter mold half may expand more than the cooler half, so that the guide pins no longer line up with the guide-pin bushings, resulting in galling or binding as the mold closes.

If a surface pyrometer is not available for measuring mold surface temperature, the latter can be estimated by touch; one's hand can be held indefinitely on metal at temperatures up to about 130°F. Above 130–140°F, the mold is uncomfortably warm to the hand, so that manual contact cannot be sustained for more than a few seconds.

CAUTION: Be sure the safety gate is open and the machine motor shut off before inserting the hand into the open mold!

Injection Pressure

It is preferable to operate at maximum available injection pressure, simultaneously reducing cylinder temperature to achieve faster molding cycles. Good practice is to increase pressure gradually as cycle is reduced, rather than to increase cylinder temperature. With high stock temperature, excessive pressure may cause flashing, or the mold may "lock up" so that it cannot be opened by the hydraulic pull-back on the movable platen. With low stock temperature, excessive pressure may cause:

1. Damage to the torpedo (spreader) on conventional straight plunger machines, or
2. High molded-in stresses at the gate area.

If the machine is so equipped, it is preferable to use a higher "first-injection" pressure while the cavities are filling and a lower "injection-holding" pressure to prevent back-flow out of the cavities after they are filled.

On such machines, the high first-injection pressure should be set so as to drop to the lower "holding" pressure as soon as the cavities are filled, which is the instant at which the plunger stops moving forward (see Plunger Forward Time, below).

Feed Cushion

The feed cushion is the difference in final plunger position from its maximum forward position during the injection stroke. At zero feed cushion, the plunger "bottoms out" by reaching the mechanical limit of its forward stroke, at which point no injection pressure is being applied to the plastic melt in the cylinder. A high feed cushion requires the plunger to push against more material in the cylinder to cause movement through the nozzle. This causes a greater pressure loss from plunger face to nozzle.

As a general rule, feed should be adjusted so as to give minimum cushion, preferably about ⅛ in. as shown on the scale of plunger position. When molding regrind or nonuniform particle sizes, a higher feed cushion (½ to ¾ in.) will help eliminate trapped air. If the material contains moisture or excessive volatiles (as shown by black streaking in the molded parts), it is frequently helpful to increase feed cushion to about 1 in., simultaneously reducing the rear cylinder zone temperature setting to about 30°F below the center zone setting.

Weigh feeders eliminate the need for careful adjustment of volumetric feed control and insure uniform (and minimum) cushion once the feed weight is adjusted to equal the shot weight.

Plunger Forward Time

Plunger forward time is the interval from the instant the plunger begins its forward stroke to the instant the injection pressure is released. The timer controlling plunger forward time is actuated at the instant the plunger starts its forward injection stroke; at the expiration of the set time the plunger starts to retract.

On one-stage straight plunger machines and reciprocating in-line screw machines, it will be observed that during the injection stroke the initial plunger speed suddenly drops to a much lower speed as it approaches its ultimate forward position. Usually the initial speed will be at the rate of 1 in. per sec or more as the mold cavities are filling; this speed drops to less than 1/16 in. per sec rather suddenly as soon as the cavities are filled, and further plunger travel acts to "pack" the cavities.

The initial high-speed portion of the plunger stroke is the *fill time* or *plunger travel time,* after which the plunger creeps forward slowly until the gates freeze.

If the plunger forward timer is set to time out at the instant the cavities are filled (i.e., at the end of the plunger travel time), injection pressure stops immediately and plastic begins to flow *out* of the cavities until the gates freeze. If the plunger begins to retract immediately after the fill time, it also tends to create a negative pressure or suction within the cylinder. Under these conditions, the molded article will show sink marks, orange peel, or dogskin. Accordingly, it is necessary to maintain a positive pressure for a certain period of time after the cavities are filled until the gates freeze. Ideally. a high injection pressure should be maintained during fill time, dropping immediately to a lower holding pressure that is maintained until gate freeze-off.

Recommended practice is to reduce progressively the plunger forward timer setting (keeping the overall cycle constant) until sink marks begin to show on the molded parts. When this point is reached, the plunger forward timer should be increased in steps of 1 sec every three or four cycles until sink marks or other low pressure defects disappear. This procedure establishes the minimum plunger forward time required for gate freeze-off at the prevailing stock temperature and mold temperature. Maintaining this plunger forward timer setting, the overall cycle can then be progressively reduced, until warpage or distortion begins to occur on ejection.

Molding Cycles

Molding cycle, or overall cycle time, is the total time from the instant the dies close during one cycle to the corresponding instant of die-closing on the next cycle while the machine is operating on a repetitive production basis. The term *machine cycle* refers to that portion of the overall cycle time that is controlled by the preset machine cycle-element timers. The machine cycle begins when the operator closes the safety gate, which activates the mold-closing stroke. Subsequent operations of mold closing, injection stroke, cooling time, and mold-opening stroke occur automatically. The machine cycle ends when the "Dies Closed" timer times out and the machine opens to the limit of the preset mold-opening stroke. The overall cycle time is thus the sum of the machine cycle time and the time required for the operator to open the safety gate, remove the molded shot, and close the safety gate.

Machine cycle time is the cumulative total of the following: the time for injection; the time required to cool the molded article to a rigid state; the time element for the mold-opening stroke; and the time element for the mold-closing stroke. The last two time elements are operating characteristics of the machine itself and are essentially independent of the adjustable molding variables and the polymer properties. The cooling time is the longest portion of the machine cycle time and is governed primarily by the thickness of the molded article, the mold temperature, and the "set-up" characteristics of the polymer.

A common desire is for a fast cycle. The hourly operating cost of a given injection machine is a constant, so that a faster cycle yields a lower molding cost per piece. However, excessively fast cycles can cause more rejects and therefore fewer acceptable pieces per hour. Rejected pieces also represent a definite material value, and even if these are salvaged by grinding and molding the regrind, the salvage cost contributes an added cost per piece. Furthermore, extremely short cycles impose an increased burden on the hydraulic system of the machine, which ultimately means greater downtime and increased maintenance expense.

In some cases, refrigerated water is used to chill the molds in order to achieve faster cycles. Circulation of cold water through the mold is not objectionable, provided that the cavity surface temperatures are not below ambient room temperature. Such low surface temperatures produce highly strained, brittle moldings having poor surface finish. Also, when mold temperature is below the dew point of the air, moisture condenses on the mold and causes rusting of steel mold surfaces.

As stated above, the cooling time required to solidify the plastic to a sufficiently rigid state is a function of part thickness, mold temperature, and material set-up characteristic. Typical machine cycle time as a function of average part thickness is shown below:

Thickness (in.)	Machine Cycle (sec)
0.020	10
0.040	15
0.060	22
0.080	28
0.100	35
0.120	45
0.140	65
0.160	85

For wall thickness greater than about 0.150 in., it is usually only necessary to chill the outermost layer of the plastic piece to a sufficiently rigid state to permit ejection. This will be governed by the size and number of ejector pins bearing on the molded piece. If the total area of ejector-pin bearing surface is relatively large, there is little danger of the concentrated force exerted by the ejector pins causing distortion of the molded piece, which can occur if the ejector pins are small diameter and few in number.

The following techniques can be used to achieve faster cycles:

1. Preheat the molding material in the hopper by means of a hopper dryer, delivering air at 180–200°F. This heat input will permit reduction of cylinder zone temperature settings, thus yielding faster set-up in the mold.
2. Adjust material feed setting to provide minimum feed cushion. This reduces pressure loss in the cylinder, yielding faster injection rates at lower cylinder temperatures.
3. Polish all cavities to remove undercuts, and provide sufficient taper, so that the use of operator-applied release agents on the molds are not required. Each application of the mold release agent requires several seconds, so that cumulative time loss can amount to several minutes per hour.

4. Control mold temperatures so that the die that forms the appearance surface of the molded article is 30–50°F hotter than the die that forms the nonappearance surface.

Example: Maintain cavity surface temperature at 140°F, and force surface temperature at 100°F.

5. On straight plunger machines, adjust the prepositioning limit switches so that after plunger retracts fully to pick up feed charge for the next cycle, it advances a partial stroke before the mold opens.
6. On moldings having section thickness over 3/16 in., immerse the molded shot into a warm water bath immediately after ejection from the mold.
7. Use shrink fixtures (preferably wood or felt-covered metal) to clamp the parts immediately after ejection.
8. Maintain a log sheet or operating instruction card for each mold and each machine. This will facilitate mold start-up and minimize lost time, reaching optimum conditions on repeat production runs.
9. Grind sprues, runners, and short shots immediately, and feed this regrind back to the machine as fast as it is generated. The value of one pound of regrind material at 20¢/lb is equal to 1 min of molding time on a machine whose operating cost is $12 per hour.

Molding Shrinkage

Mold shrinkage is the difference between the size of the molded article after it reaches equilibrium at room temperature and the size of the cavity in which it was molded.

Mold shrinkage is influenced by mold design and molding conditions. Any factor that tends to *increase* effective pressure within the cavity will *reduce* the shrinkage and yield a larger size part.

To reduce shrinkage:

1. Increase plastic stock temperature
2. Increase injection pressure
3. Reduce feed cushion
4. Reduce mold temperature
5. Increase gate size or gate land

6. Increase nozzle size
7. Install multiple gates
8. Increase plunger forward time
9. Increase ram speed
10. Increase mold-closed time
11. Increase cavity venting

To *increase* shrinkage, employ procedures opposite to those listed above.

Dimensional Tolerances

As described above, molding conditions and mold-design factors affect the molding shrinkage. Even under the best conditions, there will be some small variations in stock temperature, injection pressure, mold temperature, feed cushion, and overall cycle time. For example, the cylinder zone temperature controllers will cycle off and on; the injection pressure tends to drop as the hydraulic oil heats up after prolonged operation; feed cushion may vary as particle size varies; and mold temperature will fluctuate. Also, some variation in flow properties can be expected between different lots of polymer.

Commercial molding tolerances that should be attainable with most molds are ±0.002 in. for dimensions up to 1 in., plus an additional ±0.001 in. for each additional inch. For example, a dimension of 12.340 in. can be held to tolerances of ±0.013 in.

By careful control of all molding variables, more precise tolerances can often be maintained when absolutely required. These precision tolerances are ±0.001 in. for dimensions up to 1 in., plus an additional ±0.0005 in. for each additional inch. For example, precision tolerances on a 12.340 in. dimension would be ±0.0065 in. It should be noted that maintaining these precision molding tolerances generally involves increased costs on the order of 10 to 20% higher than for holding standard commercial tolerances.

SCREW PLASTICIZATION IN INJECTION MOLDING

Prior to the advent of screw plasticization in the years 1955–1960, injection molding machines employed simple reciprocating plungers to force the plastic granules from the hopper discharge through the heating cylinder and into the mold. In the original single-plunger machines, the plastic material was propelled in stages through the cylinder by successive strokes of the plunger, becoming progressively softer and more fluid as it advanced to the injection nozzle. The pressure exerted by the plunger to force molten polymer through the nozzle and into the mold had to be transmitted through a relatively large mass of unfused and partially softened granules in the rear zone. The large pressure drop inherent in this system limited the practical shot capacity to about 48 oz.

With the growing demand for machines capable of injecting larger amounts of plastic with a single stroke of the plunger, the machine builders added a separate *preplasticizing* cylinder to the main injection, or *shooting,* cylinder. The preplasticizing cylinder, with its own separate reciprocating plunger, was mounted above the injection cylinder, so these two-stage machines were also called piggyback machines. However, even these two-stage designs did not provide sufficient heating capacity to soften plastic materials rapidly and uniformly to the desired molten state. Consequently, the rotating screw was adopted from the screw extruder, to provide increased output of uniformly heated polymer melt.

Screw plasticization (or plastication) refers to the use of an extruder screw, rotating intermittently in the injection cylinder, either as:

1. A nonreciprocating first stage in a two-stage preplasticizing machine, or
2. The combination screw-ram in a single-stage reciprocating screw-ram machine.

The principal feature of screw plasticization in either type machine is the ability of the screw to "melt" polymers rapidly and uniformly, with thorough homogenization of both temperature and composition. A well-designed screw unit produces a melt having uniform viscosity and temperature at high output rate with minimum thermal degradation. This basic feature of screw injection equipment provides the following advantages as compared to straight plunger machines:

1. Improved ability to plasticize properly those polymers having a high melt viscosity
2. Greater physical and temperature uniformity in the melt, thus providing:
   a. Better surface appearance and gloss
   b. Lower injection pressure requirements

   c. Less warpage and better control of dimensional tolerances
   d. Lower molded-in strains
3. Improved processing of scrap and regrind
4. Better color dispersion in dry-colored or concentrate-colored materials
5. Less sensitivity to variations in material particle size
6. No pressure loss in precompressing pellets or granules in the rear zone
7. Rapid and more efficient clean-out in changing from one type of plastic to another (particularly true of reciprocating screw machines)
8. Better devolatilization characteristics and greater ability to remove undesirable volatile matter
9. Frequently, reduction in overall cycle time as a result of faster injection rates and/or higher plasticizing capacity

Even with these distinct advantages, it is still possible to produce highly stressed or other poor quality parts if the screw machine is operated above its capacity, or on a too-short cycle.

The same principles that apply to the molding process in straight-ram and two-stage plunger machines also apply to screw injection equipment. Molding variables other than screw operation are the same as for conventional plunger machines and are governed by the requirements of the mold.

Screw Design

General Screw Geometry. Design characteristics such as pitch, helix angle, and length of zone sections are generally adapted from conventional extruders. The predominant design is the constant-pitch, variable-root geometry, in which the screw pitch or lead is about equal to the major diameter. With this type screw, current design practice tends to divide the screw into three distinct zones in the following order, beginning at the hopper end of the injection machine: feed, transition, and metering zones.

The function of the feed section, which consists of several flights having a fairly deep and uniform depth, is to deliver to the transition section the quantity of pellets or granules required to keep the discharge end of the screw completely filled with molten polymer. Compression, melting, and mixing are then initiated and intensified in the transition section;

flight depth in this section generally decreases uniformly from the end of the feed section to the beginning of the metering section. The metering section typically consists of several flights having a comparatively shallow, uniform flight depth. Its function is to complete the melting and mixing of the polymer so that maximum homogeneity, both physically and according to temperature, is attained.

Lengths of 8 to 10 screw diameters for the feed section, 4 to 6 diameters for the transition section, and 4 to 6 diameters for the metering section are typical of current practice.

Due to the wide difference in rheological behavior of the various commercial thermoplastics, *there is no single screw design that performs equally well with all materials.* Consequently, each manufacturer of screw injection equipment has developed a so-called general-purpose screw as a design compromise to process a wide variety of commercial polymers. As more technology is developed, improvements in screw design can be expected, which will provide better performance.

Compression Ratio. This term refers to the ratio of the volume of a metering section flight. Compression ratios available in current commercial equipment are in the range of 1.5–4.5 to 1, with the general-purpose screw usually having a compression ratio of 2.5–3.0 to 1.

While compression ratio itself can be less important in defining screw performance than the actual metering flight depth, generally speaking, screws having a high compression ratio should be operated at slower screw speeds (e.g., 10–50 rpm), while screws having a compression ratio less than 3:1 can be operated at higher screw speeds (50–150 rpm).

Screw Length. Screw length is commonly expressed in terms of the L/D ratio, which is the ratio of the effective screw length (L) to the major screw diameter (D). The L/D ratio ranges from 12/1 to 24/1 on commercial injection screws. Reciprocating screws having an L/D ratio greater than 20/1 are more difficult to support properly for concentric rotation, and therefore may be prone to excessive wear. However, because a large L/D ratio provides delivery of a more uniform melt, L/D ratios in the 18/1 to 20/1 range are most commonly used.

On the majority of reciprocating screw machines, the *effective* L/D ratio diminishes with length of injection stroke. This reduction in effective L/D ratio is only two or three units, however, and is therefore less significant with high L/D ratio screws. The higher the L/D ratio, the less screw performance is sensitive to variations in material particle size.

Flight Depth. Generally speaking, the greater the flight depth, the greater the delivery rate in pounds per hour output at a given screw speed. Current design practice is to use a shallow flight depth for less viscous materials and deeper flights for polymers having a high melt viscosity. However, a too-great flight depth can cause torque overloads and screw stalling. Commercial designs have a flight depth ranging from 0.15 to 0.18 D in the feed section and 0.05 to 0.06 D in the metering section. Because of the intermittent operation of injection screws, flight depth is less critical than in conventional extruders in which the screw rotates continuously.

Screw Drive Characteristics

The screw drive unit should provide a high torque output with protection against torque overload and make provision for variable screw speed. The drive for an injection screw differs from that of a conventional extruder screw in that the injection screw must start and stop many times per hour, cycling with the repetitive molding cycle. Variable screw speed is necessary to accommodate different plastic materials and the variety of molds that may be operated.

In electric drives, screw speed is adjusted in steps by changing gears. Screw speed is a constant independent of load, whereas torque varies with applied load.

Hydraulic drives employ hydraulic pumps and gear reducers and provide constant torque and effective overload protection. Screw speed can be continuously variable by stepless adjustment, but it may vary with applied load.

Nonreturn Valves

A reciprocating screw machine usually has a nonreturn valve attached to the front end of the screw to prevent the melt from flowing back along the screw flights during the injection stroke, when the screw is functioning as a nonrotating plunger. A nonreturn valve is required when:

1. The melt viscosity is low.
2. The injection pressure is high.
3. The L/D ratio of the screw is low.

Nonreturn valves are commonly of two types: those having an external check ring, and those having an internal ball check. Both types can cause hang-up, with resultant thermal degradation and black streaking, and therefore should be examined periodically. Check ring valves usually exhibit some wear, necessitating replacement after a period of time. Excessive wear or improper operation of the check ring valve can be an unsuspected source of degradation or black streaking. If black streaking persists after thorough cleaning of the screw and valve, and other possible sources of streaking have been investigated, the machine manufacturer should be notified.

Cylinder Temperature Control

The cylinder zone temperature controllers on any screw injection machine usually do *not* indicate the true temperature of the melt leaving the screw. Because of frictional heat generated by screw rotation in shearing the melt, the melt temperature is frequently considerably higher than the temperatures set on the cylinder zone heater controllers. The center and front zone controllers frequently override, even with these heaters shut off. This heater override will usually occur if the screw rotation time is more than half the overall cycle time, or if high screw speeds are used, or if excessive screw back pressure is used (see below). Any of these conditions can cause excessive frictional heating of the polymer, such that the temperature of the melt is no longer controlled by the cylinder zone heater controllers.

If excessive frictional heating is caused by prolonged time of screw rotation during the screw-ram retraction stroke, this problem can be minimized by:

1. Increasing the screw speed to a higher rpm
2. Increasing the overall molding cycle
3. Reducing screw back pressure
4. Increasing rear zone cylinder temperature
5. Use of molding material having less external lubricant

Proper zone temperature settings depend primarily on the type of polymer being processed. As is the case with conventional plunger machines, it is the actual melt temperature that is important. Materials having relatively high melt viscosities and high impact require that the temperature of the rear or feed zone be set fairly high, particularly if the material is not preheated, or if the molded shot weight is close to the rated machine capacity.

If the molded shot weight exceeds 75% of the rated machine capacity (expressed in ounces of polystyrene), or if the material is not preheated, the rear zone controller temperature should be set 30–50°F higher than the center and front zones. Where the molded shot weight is between 50% and 75% of the rated machine capacity, all zone controllers should be set at the same temperature. If the molded shot weight is less than 50% of the rated machine capacity, the rear zone controller should be set 30–50°F lower than the center and front zones.

Recommended practice is to have the screw frictionally generate as much of the total heat input as possible, at the lowest practical cylinder zone settings consistent with proper control of melt temperature. Melt temperature should be measured by pyrometer needle probe of a nozzle air shot taken while the machine has been on cycle at equilibrium operating conditions, taking proper safety precautions. Installation of a thermocouple in the plastic melt stream at or near the nozzle to indicate melt temperature continuously is a distinct advantage.

Screw Back Pressure

Back pressure is that pressure that the screw must develop in the front cylinder zone to pump the melt forward. Back pressure is varied by adjusting the setting of the relief valve on the discharge line from the hydraulic injection cylinder.

On reciprocating screw machines, the back pressure must be overcome by the pumping action of the screw as it rotates to feed the charge for the next injection stroke. On two-stage machines, it is the pressure that must be overcome by rotation of the nonreciprocating screw in the preplasticizer cylinder to force the injection plunger back to its retracted position in preparation for the next injection stroke.

Increasing back pressure increases melt temperature and also increases screw rotation time. Higher back pressure intensifies shearing, thus promoting better melt uniformity, better color mixing, and more consistent moldability. Back pressure, however, should be kept to a minimum with polymers having a low melt viscosity in order to avoid backflow along the flights as well as a reduced pumping rate. In the case of polymers that have relatively high melt viscosities, increasing screw back pressure gives parts having better gloss and lower molded-in stresses. However, excessive back pressure can cause thermal degradation, discoloration, or splay marks.

Screw Speed

Plasticizing capacity of screw injection machines is controlled primarily by screw speed. The higher the speed, the greater the output rate. The frictional heat generated by screw rotation is approximately proportional to the square of the speed. However, increasing the screw speed reduces the total time of screw rotation during the molding, thereby reducing the indicated temperature shown on the cylinder pyrometer.

Optimum screw speed is a function of the screw design characteristics (compression ratio and L/D ratio) as well as of the rheological properties of the polymer involved. The melt temperature of the higher viscosity resins is particularly sensitive to changes in screw speed. Excessive screw speed can cause thermal degradation, resulting in dark streaks, splay marks, and weaker molded parts (i.e., higher molded-in strains). However, high screw speeds insure a fast recharge rate and a more uniform melt, thus permitting faster cycles.

Recommended practice is to adjust the screw speed high enough so that the time of screw rotation is about one-third of the total overall molding cycle time. It is preferable to time the screw rotation to stop just before the mold opens. If the screw rotation stops earlier, the polymer melt is held at high temperature longer and is more prone to scorching or thermal degradation.

Rate of Injection

Screw injection equipment is capable of faster injection rates than conventional plunger machines because there is no pressure loss in precompressing unfused granular or pelleted material in the rear cylinder zone, nor is there a torpedo or spreader impeding pressure exerted by the plunger. In plunger machines, the pressure drop in compressing unmelted polymer particles in the feed section can be as much as 50% of the total pressure exerted at the face of the plunger. Accordingly, in screw injection machines, a much lower plunger pressure is required to develop the same injection pressure *at the nozzle* than is the case with plunger machines. Note that high injection pressure *at the nozzle* is still necessary on all types of machines to attain a fast injection rate.

High rates of injection are generally beneficial in reducing molded-in orientation stresses and promoting shorter molding cycles, and are particularly advantageous for long-flow parts having wall thicknesses of less than ⅛ in. However, undersize gates or nozzles impose a limit on injection rate. Excessive injection speed can cause jetting, frictional burning, or delamination in the gate area. Similarly, heavy-section parts such as brush backs and shoe heels require slower injection rates to avoid flow marks and other surface defects.

Because screw injection machines are usually capable of faster injection rates than plunger machines, control of this variable is more critical in screw equipment. Rate of injection can be reduced by decreasing the melt temperature, and so screw machines are capable of shorter molding cycles when molding heavy-section parts, because the lower stock temperature coupled with more uniform melt permits faster set-up in the mold.

Recommended practice is to use high injection pressure and as fast a fill rate as the part and mold design will tolerate, and immediately drop to a lower holding or dwell pressure as soon as the cavities are filled. This practice will minimize molded-in stresses. Except for heavy-section parts (i.e., section thickness greater than ¼ in. or so), the fill time or plunger travel time should not exceed 5 or 6 sec. Obviously, fill time depends on the size, number, and location of gates, the extent of mold-cavity venting, the nozzle size, and the flow properties of the polymer.

INJECTION MOLD DESIGN

The injection mold is the heart of the injection molding process, for it is the tool that forms the molten plastic into the desired shape, provides the surface texture, and determines the dimensions of the finished molded article. The mold also influences the internal stress level of the molded object, and therefore has a significant effect on properties and end-use performance.

An injection mold is a precision instrument whose cost can range up to $100,000, and yet it must be rugged enough to withstand hundreds of thousands of high-pressure molding cycles. It has been said that 90% of the success of any molding job depends on the skills employed in the design and construction of the mold. Certainly, it is impossible to produce satisfactory moldings in a poorly built mold. Proper design and skilled craftsmanship in mold construction greatly facilitate economical production of articles exhibiting the maximum performance of which the plastic material is capable.

Mold making is still largely an art rather than a science. Imagination, knowledge, and experience are of paramount importance in successful mold design. The information in this chapter, combined with these subjective factors, will serve as a useful guide in designing molds.

Basic Design Considerations

The basic mold design is governed by:

1. The size and shape of the molded article
2. The number of cavities to be installed in the mold
3. The size and capacity of the injection machine in which the mold is to be operated

These factors are interrelated, but the size and weight of the molded object limits the number of cavities in the mold and also determines the machine capacity required. In the case of a large molded object, such as an auto radiator grille or a one-piece bucket chair, the large exterior dimensions of a single-cavity mold require a correspondingly large clearance between the machine tie-rods. Similarly, the machine tie-rod clearance limits the number of cavities that can be installed in a multicavity mold.

The total projected area of the molded shot determines the clamping force required to hold the mold closed during the injection stroke. The projected area can be defined as the area of the shadow cast by the molded shot when it is held under a light source, with the shadow falling on a plane surface parallel to the parting line. Note that the projected area includes the area of the runners (except in a hot runner mold).

For a true hydraulic fluid such as water, the clamping force required for each square inch of projected area would be equal to the unit pressure applied by the injection plunger. However, due to the partial hardening of the plastic as it flows through the sprue, runners, and into the cavity, the actual pressure exerted by the plastic within the cavity is much less than the applied plunger pressure. For this reason, an applied plunger pressure of 20,000 psi would seldom require a clamping force of more than 5 or 6 tons per sq in. of projected area of the plastic shot.

For a given plunger pressure, the actual pressure developed within the cavity varies directly with the thickness of the molded section, and varies inversely with the melt viscosity. Thick sections require greater clamping force than thin sections because the plastic melt in a thick section stays semifluid for a longer time during the cavity-filling injection stroke. Similarly, a higher stock temperature, a hotter mold, larger gates, or a faster rate of injection require a higher clamping pressure. As a general rule, good molding practice requires about 3 tons of clamp for each square inch of projected area of the molded shot.

By proper mold design and careful adjustment of molding conditions, it is sometimes possible to mold satisfactory parts with as little as 1 ton of clamp per square inch of projected area. However, it is unwise to attempt to operate a mold on this basis, as the range of permissible molding conditions would be seriously limited, and long-flow cavity fill could not be achieved.

It is also important to avoid applying too much clamping force to the mold. If a small mold is installed in a large machine and closed under full clamp, the mold can actually sink into the machine platens. Also, if there is insufficient area of mold steel in contact at the parting line, the mold may be crushed under the excessively high clamping force. Steel molds will begin to crush when the unit clamp pressure exceeds 10 tons per sq in. of contact area. In severe cases, the mold components may distort, or fracture prematurely from fatigue.

While the projected area determines the clamping force required, the weight of the molded shot determines the capacity of the injection machine in which the mold must be operated. Note that the shot weight includes the weight of the sprue and runners, except in hot runner molds.

Capacities of injection machines are commonly rated in ounces of polystyrene that can be injected by one full stroke of the injection plunger. A better capacity rating, however, is the volume of plastic displaced by a full stroke of the plunger. For a given machine, this volumetric displacement is a constant that is independent of the specific gravity of the plastic. For example, a machine having a rated capacity of 48 oz of polystyrene has a plunger displacement of 78.7 cu in. because 1 cu in. of polystyrene weighs 0.61 oz. However, this same machine will inject 63 oz of rigid vinyl (0.80 oz per cu in.) in one stroke, but only 41.7 oz of polyethylene (0.53 oz per cu in.).

All too frequently, a mold is operated at or near the maximum rated capacity of the machine, or even in excess of this capacity. The inevitable result is poor quality: sink marks, heavy flow lines, poor welds, dull finish, and erratic production. Operating at full rated capacity usually necessitates a longer molding cycle to allow uniform plasticization of the materials to the desired melt temperature. Furthermore, prolonged operation of the machine at its rated capacity imposes a heavy burden on the heating cylinder and hydraulic system, resulting in premature failure of the overloaded heating bands, pumps, and valves.

On two-stage (preplasticizer) plunger machines or reciprocating screw machines, a higher percentage utilization of rated machine capacity is practical, but it is still preferable to operate below 90% of the rated capacity.

If circumstances require that the molded shot weight exceed these recommended percentages, the following procedures are suggested:

1. Preheat the plastic molding material for 2–3 hours at 170–190°F.
2. Set the rear-cylinder-zone temperature 30–50°F higher than the center and front zones.
3. Lengthen the overall cycle time by 5–10%.
4. On screw-ram machines, increase the screw back pressure to 100–500 psi.
5. Use a softer-flowing grade of plastic or a higher level of external lubricant.

Functions of the Mold Components

The injection cylinder converts the plastic granules to a molten state, and the plunger delivers this plastic melt through the nozzle under high

Mold Component	Function
1. Mold base	Hold cavity (cavities) in fixed, correct position relative to machine nozzle
2. Guide pins	Maintain proper alignment of two halves of mold
3. Sprue bushing (sprue)	Provide means of entry into mold interior
4. Runners	Convey molten plastic from sprue to cavities
5. Gates	Control flow into cavities
6. Cavity (female) and force (male)	Control size, shape, and surface texture of molded article
7. Water channels	Control temperature of mold surfaces to chill plastic to rigid state
8. Side (actuated by cams, gears, or hydraulic cylinders)	Form side holes, slots, undercuts, threaded sections
9. Vents	Allow escape of trapped air and gas
10. Ejector mechanism (pins, blades, stripper plate)	Eject rigid molded article from cavity or force
11. Ejector return pins	Return ejector pins to retracted position as mold closes for next cycle

pressure. As the plastic leaves the nozzle, the mold must perform all functions. The importance of mold design is emphasized by the fact that good molded parts cannot be produced if any *one* of these functions is not performed properly.

Sprue Bushing

The front end of the sprue bushing contains a spherical depression or recess to fit the cylinder nozzle. The sprue channel is tapered to diverge from the nozzle in order to permit easy release of the plastic sprue. Standard taper should be ½ inch per ft. The orifice at the nozzle end of the sprue will separate readily from the nozzle when the mold opens. The sprue orifices should be at least ¼ in. in diameter. The length of the sprue channel should be as short as possible and never longer than 4 in. Where mold-design factors indicate the need for a longer sprue, it is good design practice to recess the sprue bushing deeper into the mold, and install an extended heated nozzle on the injection cylinder. The sprue channel should be highly polished and free from all burrs, tool marks, or undercuts.

In single-cavity molds, the sprue usually enters directly into the cavity, in which case the sprue diameter at the point of cavity entry should be approximately twice the thickness of the molded article at that point. Insufficient diameter of the direct sprue gate can cause excessive frictional heating and/or delamination of the plastic at the gate area. A sprue diameter that is too large requires a prolonged molding cycle to allow the plastic sprue sufficient time to cool for removal. In all direct-sprue-gated cavities, an internal water "fountain" should be installed in the mold directly opposite the gate to cool the mold surface. All plastic injected into the mold impinges on this surface and causes a "hot spot" on the metal cavity wall.

In three-plate and hot runner molds, the main sprue is designed as described above. The smaller sprues (also known as sub-sprues), which convey plastic from the runners to the cavities in such molds, are designed to converge toward the gates.

Runners

The runners in multicavity molds must be large enough to convey the plastic melt rapidly to the gates without excessive chilling by the relatively cool mold. Runner cross sections that are too small require higher injection pressure and longer time to fill the cavities. Large runners produce a better finish on the molded parts and minimize weld lines, flow lines, sink marks, and internal stresses. However, excessively large runners should be avoided, for the following reasons:

1. Large runners require longer to chill, thus prolonging the molding cycle.
2. The increased weight of a large runner system subtracts from the available machine capacity, not only in terms of ounces per stroke that can be injected into the cavities, but also in plasticizing capacity of the heating cylinder in pounds per hour.
3. Large runners produce more "scrap" that must be ground and reprocessed, resulting in higher operating cost and increased possibility of contamination.
4. In two-plate molds containing more than eight cavities, the projected area of the runner system adds significantly to the projected area of the cavities, thus reducing the *effective clamping* force available.

Note that these objections do *not* apply to hot runner or runnerless molds.

A full-round (i.e., circular cross-section) runner is always preferred over any other cross-sectional shape, as it provides the minimum contact surface of the hot plastic with the cool mold. The layer of plastic in contact with the metal mold chills rapidly, so that only the material in the central core continues to flow rapidly. A full-round runner requires machining both halves of the mold, so the two semicircular portions are aligned when the mold is closed. This small additional machining cost will pay dividends in better molding performance.

Where the runner is machined in only half of the mold, the square or trapezoidal shape most closely approximates the preferred full-round runner. Half-round or flat, rectangular runner cross sections are not recommended. In the trapezoidal runner, the depth is almost equal to the width; the departure from a true square cross section results only by the need to provide ample taper on the runner sidewalls for easy mold release. To provide the recommended 5-degree taper per side, the width at the top of the runner should be 1.18 times the depth. Depth of trapezoidal runners to provide equivalent cross-sectional area to full-round runners is shown in the following table.

Diameter of Round Runner (in.)	Decimal Equivalent	Cross-Section Area (sq in.)	Depth of Equivalent Trapezoidal Runner
1/8	0.125	0.0123	0.110
5/32	0.1563	0.0192	0.138
3/16	0.1875	0.0276	0.166
7/32	0.2188	0.0376	0.194
1/4	0.250	0.0491	0.221
9/32	0.2813	0.0622	0.249
5/16	0.3125	0.0767	0.277
11/32	0.3438	0.0928	0.305
3/8	0.375	0.1104	0.332
13/32	0.4063	0.1296	0.360
7/16	0.4375	0.1503	0.388
15/32	0.4688	0.1725	0.415
1/2	0.550	0.1964	0.443
17/32	0.5313	0.2219	0.471
9/16	0.5625	0.2484	0.498

Runner diameter should be at least 1/8 in. for polystyrene and at least 1/4 in. for ABS. Longer runners require increased diameter. All main runners in a given mold should be equal in diameter, while the diameter of all secondary runners should be at least 1/32 in. less than the diameter of the main runners.

Intersections of the secondary runners with the main runner should be filletted with a 1/8-in. radius on the upstream or sprue side of the intersection. Similarly, the intersection of the sprue with the main runner should also have a 1/8-in. fillet radius.

In multicavity molds, the layout of cavities and runners should be positioned so that the flow distance from the sprue to each cavity is exactly the same. This provides equal distribution of the injection pressure to each cavity, so that the rate of flow into each cavity will be equal. This principle can be applied to any even number of cavities; however, in a six-cavity mold, it is preferable to arrange the cavities in a circular pattern, with the runners positioned in a radial pattern.

Gates

The gate provides the means for controlling the rate of flow into the cavity and also governs the degree of *packing,* or compression flow, into the cavity after it is volumetrically filled. The size and shape of the molded article determines the type, location, and size of the gate(s).

Because the gate always leaves a surface imperfection on the molded part, it should be located on a nonappearance surface, if possible. This gate-mark problem frequently prevents locating the gate in the most desirable position for best moldability, and it also requires that the gate be small to minimize or eliminate gate removal, which can involve costly buffing operations.

To minimize the distance the plastic must travel in filling the cavity, the gate should always be located as near the center of the part as design factors will permit. Where part design or appearance considerations prohibit a central gate location, a larger gate or several gates per cavity should be provided to permit more rapid fill. In the majority of cases, gate location is limited by the necessary position of the cavity and of the parting line. It is desirable to locate gates in the thickest section of the molded article, as the thick section is the slowest to chill and therefore the most prone to show sink marks. It is also preferable to locate gates at points of minimum end-use stress because the gate area is usually the weakest portion of the molding.

Flow distance refers to the length of the path the plastic must travel in flowing from the gate to the farthest extremity of the cavity. The term *flow ratio* is the ratio of the flow distance to the nominal section thickness of the molded part. Each plastic material has a limiting flow ratio, which is a characteristic of its rheological and set-up properties.

For molded articles in which the use of a single gate would result in a flow ratio exceeding these limiting values, two or more gates should be used. A typical example is a refrigerator breaker strip, which may have an overall length of 36 in. and a section thickness of 0.060 in. With a single gate located at the midpoint of one long edge, the flow ratio would be 18 in. divided by 0.060 in., or 300:1. In this case, the mold could not be filled with high-impact polystyrene, unless a long flash gate were used. Cavity fill-out could also be insured by using two or more gates to reduce the flow distance. The use of two gates always causes a weld line to form where the two streams of plastic come together. The prominence and weakness of this weld increases as the flow distance from the gate to the weld increases. Accordingly, it is better to use three or four gates, as each weld line will then be stronger and also less discernible. Multiple gating also permits each gate to be thinner, and therefore gate freeze-off will occur faster, enabling faster molding cycles.

In a single-cavity mold, the cavity should be positioned symmetrically with the nozzle axis in order to insure uniform distribution of injection and clamping pressures. For the same reason, in a multicavity mold, the cavities must be positioned symmetrically with respect to the sprue. These requirements frequently dictate gate location.

The large, round gate and the similar half-round and square gates were the types first used in early mold designs. These relatively large gates offer the minimum resistance to flow, and were used in the early days of injection molding because of the limited injection pressure available, and also because early thermoplastics were either heat-sensitive or stiff-flowing. Such large gates are still used extensively today for molding heavy sections, such as brushback and for molding stiff-flowing or heat-sensitive polymers. Generally, the large round gate is used for parts whose thickness exceeds ¼ in., such as brushbacks and thick knobs. Large gates are desirable for minimizing orientation stresses in molding, but they also require a longer time to chill in the mold. This results in packing, caused by the high pressure in the runner that forces more plastic into the cavity after it is filled, resulting in high internal stresses in the gate area.

The *fan gate* is simply a flattened version of the large round gate and is used for parts having large areas and relatively thin wall sections, such as clock faces, speedometer dials, and similar parts with a high ratio of surface area to thickness. The thickness of the fan gate should be no more than half the wall thickness of the molding, so that the gate will freeze rapidly after the cavity is filled to prevent the packing stresses mentioned earlier. The width of the fan gate is governed by the size of the cavity, the flow pattern within the cavity, and the rate of filling required. Unfortunately, there is no way to predict the width needed. The most practical approach is to start with a gate thickness measuring one-third of the part thickness and with a width of ⅛ in., and then progressively widen the gate until the desired flow pattern is achieved. This may seem a tedious and time-consuming process, but it pays dividends because thin, small gates permit the fastest molding cycles.

The *gate land,* or length of the gate between the runner and the cavity, should be short, preferably 0.020 to 0.030 in. If less than 0.020 in., the steel wall that forms the land may not withstand the high pressure in the runner, or the gate may be difficult to trim. A gate land length over 0.030 in. causes extreme resistance to flow through a thin gate, necessitating higher injection pressures, higher stock temperatures, or both.

A fairly recent development, the *flash gate* is very well suited to large flat areas. A secondary runner is cut parallel to the cavity wall, fed at its midpoint by the main runner. The gate land, or distance between the cavity and the parallel runner, is 0.020 to 0.030 in. Depth of the gate is 0.020 in. but the width is several inches. The thin flash gate enables the cavity to fill rapidly, yet it chills quickly, permitting fast cycles. Note that a flash gate 0.020 in. deep by 6 in. wide has a greater cross-sectional area than a fan gate 0.050 in. deep by 1 in. wide, and hence will enable the cavity to fill faster. The disadvantage of the flash gate is the large parallel runner that must be reprocessed. Accordingly, a good compromise design is one intermediate between a fan gate and a flash gate.

Pinpoint, or restricted, gates are gaining in popularity and should be used whenever the design or size of the part permits. Pinpoint gates should not be used with highly viscous polymers, or those that are heat sensitive. The pinpoint gate is the best way of controlling flow into the cavity, and so is of great value in balancing the gating in multicavity molds. Because of its thin dimensions, the pinpoint gate chills very quickly, resulting in short cycles and minimum packing stresses. Its small size also reduces or eliminates costly gate-removal operations and permits automatic degating in three-plate molds and hot runner molds.

One disadvantage of the pinpoint gate is that the smaller the gate the higher the velocity of the plastic flowing through it. Some of the kinetic energy of high velocity is converted into heat, and, in extreme cases, this frictional heating may cause burning or thermal degradation. In less severe cases, *jetting* will occur. Jetting is the rapid extrusion of a long, threadlike stream into the cavity, the thread curling up and chilling rapidly. This coiled thread of chilled plastic lies in the cavity and is subsequently displaced by the incoming hotter plastic. Accordingly, surface flow marks and stresses are formed by the shearing of the chilled thread by the following plastic. For this reason, the pinpoint gate should always be located so that the entering stream of plastic immediately impinges on a force plug or core pin.

Cavity, Force, and Mold Cores

A review of all the considerations involved in the design and construction of cavities and forces is beyond the scope of this chapter. Several books and numerous articles have been written about the various factors involved in selection of mold steels, machining, hobbing, duplicating, engraving, etc. The chief points that will be noted here deal with mold metal, undercuts, and taper.

Mold Metal. Steel, particularly alloy tool steel, is the most common metal used for constructing cavities, forces, and movable cores. The use of beryllium-copper alloys and stainless steels is becoming popular, the former for faster mold cooling and the latter for its stain-resistance and rustproof qualities. Polished, hardened tool steel is recommended for moldings that require a mirror finish; such mold components can also be chrome-plated to maintain the polished surface. A matte or satin finish on moldings can be obtained by vapor honing or vapor blasting the cavity surfaces. Various textured surfaces, such as simulated leather grain, can also be reproduced on moldings by appropriate texturing of the mold. Any textured surface, however, tends to become smoother after several thousand molding cycles due to the slight erosive action of viscous plastic flowing with a scouring action over the textured surface.

Various soft metal alloys and aluminum have also been used for low-cost or short-run injection molds, but such molds must be operated carefully to avoid permanent bending or distortion under high injection pressures. Such molds have a useful life of only a few thousand cycles under normal operating conditions, while the hardened steel molds will usually withstand a million cycles.

Undercuts. There must be no undercuts or reverse tapers on the cavities or forces that will prevent removal of the molded parts from the mold. Undercuts can sometimes be molded in by side-action cores or movable mold components that are withdrawn before the ejector mechanism pushes the molding out of the cavity or off the force. Such side-action cores must be carefully designed to be foolproof and automatic in operation. It is sometimes less costly and less troublesome to drill side holes or machine the undercuts on molded parts as a secondary operation after molding. Frequently, a difficult or complex part can be molded as two or more sections and assembled after molding at a lower cost than integrally molding the part in one piece.

Taper (Draft). Adequate taper must be provided on all surfaces perpendicular to the parting line. It is good design practice to provide a 1-degree sidewall draft (0.0175 in./in.) when molding rigid thermoplastics having a low elongation at yield. For the more resilient polymers, a sidewall draft of ½ degree (0.008 in./in.) is sufficient. With well-built molds having draw polished, mirror-finish sidewalls, it is possible to mold these rigid plastics with as little as 0.005 in./in. sidewall taper.

Venting

Each cavity must be adequately vented in order to permit the escape of trapped air and gas that are displaced as the molten plastic enters the cavity. In effect, the cavity in a tightly clamped mold behaves as a sealed container with the gate being the only opening. Plastic cannot flow in unless the air escapes. This vital requirement is frequently overlooked, and the inevitable results are short shots, burned spots, weak welds, poor surface finish, flow marks, or slow filling of the cavities. The importance of adequate venting cannot be overemphasized; this design feature should receive as much attention as gating. Without adequate venting, it is impossible to achieve good molding performance.

Vents should be incorporated on the mold parting surface, by grinding a recess 0.002 in. deep by ¼ to ½ in. wide, extending fully from the cavity to the mold exterior. The depth of the vent should increase slightly as it diverges from the cavity so that the vent will be self-cleaning. The cavity can also be vented by providing a clearance of 0.001 to .002 in. around each knockout pin, or by grinding flats 0.002 in. deep, parallel to the long axis of the pin. A vent should be installed at every point on the parting line where a weld line occurs. For this reason, the vents will have to be ground in after initial molding tryout.

In deep-draw cavities, or at the tops of ribs where air tends to be trapped in a pocket, it is desirable to install vented knockout pins, or else vent by drilling a small hole through the mold and driving in a force-fit pin having 0.002-in. flats ground longitudinally. Such pockets have also been successfully vented by drilling a very small hole through the bottom of the cavity, connecting this to a larger diameter hole drilled from the reverse side.

The weld line formed by the intersection of plastic streams flowing around a meld core can be minimized by installation of a special enlarged vent known as an *overflow well.* In this design, the vent depth should be about one-third the thickness of the molded wall section so that plastic will be forced out through the vent into a cylindrical well. The well is fitted with a shortened knockout pin having 0.002-in. clearance in its drilled hole. Air trapped between the two advancing "wave fronts" of the plastic streams is vented out through this clearance; also, the chilled plastic on the "wave fronts" is forced onto the overflow well so that the final weld is formed by the fusion of hotter plastic material behind the wave fronts. In this manner, a stronger, less visible weld is formed. The cold slug of chilled plastic in the overflow well is ejected with the molded part by the knockout pin and is clipped off after removal from the mold. Such overflow wells, each with its own knockout pin, can also be used to avoid formation of a knockout-pin mark on the surface of the molded article. This is frequently desirable in the molding of speedometer dials, clock faces, and the like.

Water Channels

The temperature of the mold must be maintained at some constant temperature below the heat distortion point of the plastic in order to chill the plastic to a rigid state. The surface temperature across the mold face must be uniform, otherwise differential cooling strains may cause distortion of the molded part after ejection. Mold temperature is best controlled by circulating water at a controlled temperature through channels drilled in the mold.

This important function is rarely given enough consideration in construction or operation of the mold. A certain amount of heat input in Btu per lb must be supplied by the injection heating cylinder to convert the plastic to a flowable melt. A similar amount of heat must be removed from the mold article to chill it to a rigid state to permit ejection. Accordingly, it is essential to drill enough water channels in the mold and connect them to a source of water circulating at a controlled temperature.

Water channels should be at least ½ in. in diameter and should be located within 1 in. of the cavity wall. Water lines should run parallel to the shortest mold base dimension, and their center-to-center spacing should be 2–2½ in.

Force plugs and large cores should always be channeled for water circulation. In general, every core or force over a 1½ in. in diameter and over 1½ in. long should be channeled, either by the intersection of holes drilled at an angle to form an inverted V, or by installing an internal water fountain. In direct sprue-gated cavities, the stream of hot plastic entering through the gate creates a hot spot where it impinges on the mold force. This should be cooled by circulating water to within ½ in. of the mold surface.

The water circulation in each half of the mold should be separately controlled. Frequently, it is desirable to maintain the cavity (or appearance surface) at a higher temperature than the force (nonappearance surface) in order to obtain high gloss without sacrificing cycle time. In early days of injection molding, the two halves of the mold were connected in series so that the cold water entered the rear half of the mold first and then passed to the channeling in the front half, through a rubber-hose connection. In this way, the water became heated slightly by its passage through the rear half of the mold and automatically maintained the front half at a higher temperature.

When operating large molds, it is sometimes helpful to have two or more separate water channels in each mold half, the outermost channel being maintained at a higher temperature than the channel near the sprue in order to assist the plastic flow to the outer extremities of the cavity.

Whenever a mold is installed in a machine, the flow rate of water from the channel outlet should always be checked to insure that the channeling is not plugged and the circulating pump is operating properly. Flow rate should be great enough so that the outlet temperature is not more than 10°F higher than the inlet temperature. As preventive maintenance, it is advisable to reverse-flush the water channels occasionally and thereby remove rust, scale, and sediment.

On molds operating at fast cycles, or for large molded shot weights, it is sometimes desirable to circulate refrigerated water or brine through the water channels to remove heat rapidly. This practice is acceptable *only* if the surface temperature of the cavity and force is not chilled below room temperature; otherwise, atmospheric moisture will condense on the mold.

A temperature gradient across the mold face causes differential cooling rates of the plastic, resulting in thermal stresses in the molded article. For the same reason, it is unwise to maintain half of the mold at a temperature more than 40°F hotter than the other half because excessive temperature differential causes warpage.

Ejector Mechanisms

The most important considerations in the design of the ejector mechanism are:

1. The diameter of the ejector pins should be as large as the design permits.

2. There should be as many ejector pins as can be incorporated without interfering with the water channeling.
3. Ejector pins should bear uniformly on the molded part, to eject it smoothly and without distortion.

The pressure required to push the molded article out of the cavity, or to strip it off the face depends on the following factors:

1. Sidewall draft angle
2. Surface contact area
3. Sidewall polish of the mold surfaces
4. Injection pressure (or degree of packing)
5. Presence of release agents, either in the plastic material or applied to the mold surfaces

Total force required to eject the part can range up to several hundred pounds. If the knockout pins are too small in diameter, or too few in number, the pressure exerted by the face of the knockout pins can be sufficient to distort the warm plastic part. On short molding cycles, because of the time required for the plastic to chill to a sufficiently rigid state to resist these pin pressures, the concentrated pressure exerted by the face of the pins may even limit attainment of minimum molding cycles. For example, assume that a high-impact polystyrene article requires a force of 200 lb to eject it from the cavity. If there are only four knockout pins, each with a ⅛-in. diameter, each pin must exert a unit pressure of about 4,000 psi, which is greater than the shear strength of the warm plastic. In this case, the pins would simply push right through the part.

Although there is no proven method for determining in advance the number of knockout pins required, a rough rule-of-thumb is to provide 1 sq in. of pin bearing surface for every 100 sq in. of sidewall contact area. (Note: not projected area!) The advantage of larger-diameter pins is shown by the fact that to provide 1 sq in. of total bearing surface requires 82 pins each ⅛ in. in diameter, but only 20 pins each ¼ in. in diameter, and only 9 pins each ⅛ in. in diameter.

On a weld-built mold having proper sidewall draft and an adequate ejector system, there is no need to apply a mold-release agent such as zinc stearate dust or silicone oil spray. The apparent need to use such mold release agents is proof that the mold surfaces require more draw polishing or more sidewall draft. On a new mold, there may be microscopic tool marks that act as undercuts, so that external mold-release agents are sometimes needed during the breaking-in period. After long operation, the lands of the cavities at the parting line may become slightly peened over, creating a microscopic undercut lip. When this is suspected, the edges of the cavities should be stoned down and polished to restore proper sidewall draft. Applying mold-release agents to the mold is costly, especially in terms of the delay their application causes. Spraying on mold-release requires several seconds each time, thus prolonging the overall cycle by 5 to 10%.

Knockout pins must always be installed so that the pin face is flush with, or recessed 0.002 in. below, the cavity surface. A knockout pin that extends as little as 0.002 in. above the cavity surface causes a high stress concentration in the molded plastic at the sharp edge of the pin. In the interests of economy, molding cycles are usually reduced to the shortest possible time. This practice is almost always a false economy, for whatever is gained in pieces per hour is usually lost in higher reject rates or poorer quality. On short cycles, the interior of the plastic part is hotter than the outer layers that have been in direct contact with the cooler mold. Under these conditions, the sudden and concentrated pressure exerted by the knockout pins when the mold opens causes internal stresses that become frozen in as the part chills further.

MOLDING PROBLEMS

Successful injection molding requires that acceptable molded articles be produced consistently on a routine basis. In achieving this goal, problems can occur when molding any thermoplastic and are most prevalent when starting up a new mold. Problems can also occur when a mold is installed in a different machine, or when converting to a different plastic molding compound.

Defects in molded pieces can be caused by improper machine conditions, an unsatisfactory mold, or an unsuitable plastic material. All three factors—machine, mold, and material—must be considered when attempting to eliminate defects and achieve optimum molding conditions. All too frequently, the material or the molding conditions are blamed for defects, when the actual cause lies with the mold design or construction.

The following section lists the 10 most common defects encountered in injection molding practice. For each type of defect, the possible causes are tabulated under three headings—machine, mold, and material—any of which could be contributing factors to the particular problem.

For example, if the cavities are not filling out completely, possible causes could be:

1. Machine Conditions: Injection pressure may be too low, feed setting may not be sufficient, etc.
2. Mold: Temperature may be too low, gates and runners may be too small, the cavities may not be vented properly, etc.
3. Material: Flow may be too stiff, the pellets in the machine hopper may be too cold, or the particle size may be nonuniform.

The remedy for each possible cause is usually self-evident. In this particular case, the first step is to check the injection cylinder temperatures to see that these are high enough for the particular plastic material being molded. If possible, measure the stock temperature of an airshot to be sure the heating cylinder is actually heating the plastic to the necessary melt temperature. The second step is to determine if sufficient injection pressure is being applied during the injection stroke. At the same time, the mechanical feed control should be set to provide sufficient feed of material from the hopper, so that the injection plunger does not "bottom out" at the end of its forward stroke.

In a similar manner, the other possible causes listed under the "Machine" column heading should be explored. If all machine operating conditions appear to be satisfactory, the next step is to investigate each of the possible causes listed under the "Mold" column and, lastly, those possible causes in the "Material" column.

The step-by-step approach to troubleshooting an injection molding problem can usually be completed within a few minutes. However, correction of the problem may take much longer, especially if cylinder temperatures must be adjusted. Cylinder temperature adjustments should be made in small steps, 10°F or 20°F at a time, allowing at least ten or twelve machine cycles between changes, to allow the plastic melt temperature to reach equilibrium with the new cylinder zone settings.

A systematic analysis of the possible causes and application of the proper corrective measures will provide the most effective means to producing acceptable parts (see Molding Problems pp 72–75.)

BLOW MOLDING

This process is used to make plastic squeeze bottles and other hollow containers. A heat-softened thermoplastic tube called a *parison* is air blown against the mold surfaces then cooled before removal from the mold.

Modern machines, such as the one shown in Figure 3-4, can blow mold parts as large as 55-gal drums. Figure 3-5 shows blow molding machine for gallon-size milk containers. Figure 3-6 shows miscellaneous blow-molded articles. Figure 3-7 illustrates the blow-molding sequence.

Figure 3-4. Machine for blow molding 55 gallon drums. *(Courtesy of Uniloy)*

Molding Problems

Machine	Mold	Material
Poor Welds		
Stock temperature too low	Mold temperature too low	Flow too stiff
Injection pressure too low	Insufficient venting	Material setting too fast
Ram speed too slow	Vents plugged or peened shut	Insufficient lubricant
Nozzle too cold		
Nozzle bore too small	Gate(s) too small	Excessive volatile content
Inadequate machine capacity	Wrong location of gate(s)	Excessive lubricant
Cycle time too short	Excessive mold-release agent on mold	
	Piece cross section too thin	
	Sprue bushing too long	
	Sprue diameter too small	
	Runners too small	
Sticking in Sprue or Cavity		
Injection pressure too high	Improper seating of nozzle in sprue bushing	Insufficient mold release additive
Stock temperature too high	Nozzle bore larger than sprue orifice	Flow too soft
Plunger dwell too long	Sprue too long or too small	Insufficient lubricant
Nozzle bore too big	Sprue taper inadequate	
Booster set too high	Undercuts or poor polish on mold surface	
Excessive machine capacity	Improper knockout mechanism	
Screw speed too high	Inadequate sprue puller lock	
	Cavity edges preened over at parting line	
	Stationery die temperature too high	
Splay Marks, Silver Streaks, Surface Blisters		
Stock temperature too high	Mold temperature too low	Excessive moisture
Nozzle temperature too high	Mold temperature not uniform	Excessive volatile content

Rear cylinder zone temperature too high Cylinder heaters not functioning properly	Insufficient venting Gate(s) and runners too small	Material in hopper too cold Trace contamination with incompatible material Excessive regrind mixed with virgin material
Nozzle bore too small Ram speed too fast Inadequate machine capacity Cylinder not fully purged of previous material	Sprue diameter too small Oil, grease, lubricants, or water on mold	
Mold Flashing		
Injection pressure too high Inadequate clamping pressure Stock temperature too high	Poor alignment of mold halves	Flow too soft
	Flash or foreign matter on die faces or parting lines	Excessive or nonuniform lubricant
Booster set too high Gate-open time too long	Too many cavities Projected area of molding too great for machine clamping capacity Mold temperature too high Vents too deep	Pellet sizes too small or nonuniform
Cycle too long Excessive feed cushion Screw speed too high Excessive screw back pressure(s) Ram speed too fast		
Black Streaking or Color Degradation		
Stock temperature too high Rear cylinder zone too hot Internal torpedo temperature too high Burnt material hang-up in cylinder or nozzle seat	Grease or oil on cavities Grease bleeding from knockout pins Hot runner block too hot	Excessive volatile material Excessive fines content Excessive or nonuniform lubricant Inadequate lubricant
Nozzle improperly seated Cylinder or torpedo cracked. Worn or galled plunger		Excessive regrinds mixed with virgin material

Molding Problems (continued)

Machine	Mold	Material
Hang-up in internal check valve		
Cylinder heater band out of control (overriding)		
Oil leaking into cylinder		
Excessive Warping or Shrinkage		
Stock temperature too high	Insufficient taper or draft	Flow too soft
Stock temperature too low	Nonuniform release from core	Excessive or nonuniform surface lubricant
Plunger dwell too short	Knockouts engaging too fast, or not uniformly	Material setting too slow
Cycle time too short	Knockout pins bearing area too small	
Machine opening too fast or engaging knockouts too fast	Mold temperature nonuniform	
	Mold temperature too cold	
	Mold temperature too hot	
	Nonuniform cross-sectional thickness of piece	
	Improper side-core action	
	Core pins bent	
Short Shots (Nonfills)		
Injection pressure too low	Mold temperature too low	Flow too stiff
Stock temperature too low	Gate(s) too small	Material too fast setting
Excessive feed cushion	Insufficient venting	Inadequate lubrication
Insufficient feed	Piece cross section too thin	Too much regrind mixed with virgin material
Nozzle too small	Cold slug clogs gates	Cold material in hopper
Inadequate machine capacity	Sprue bushing too long	Nonuniform particle size
Nozzle or cylinder plugged	Sprue diameter too small	

Ram speed too low	Runners too small	
Rear cylinder zone too low		
Booster set too low		
Poor Gloss		
Injection pressure too low	Mold temperature too low	Flow too stiff
Stock temperature too low	Gate(s) too small	Inadequate lubrication
Excessive feed cushion	Insufficient venting	Cold material in hopper
Ram speed too slow	Abrupt change in section thickness	Nonuniform particle size
Nozzle bore too small		
Plunger dwell time short	Rib thickness too great	
Booster set too low		
Burn Spots		
Injection pressure too high	Insufficient venting	Flow too soft
Ram speed too fast	Vents plugged or peened shut	Excessive lubricant
Stock temperature too high	Wrong location or type of gate	Highly volatile content
Nozzle too hot	Excess mold-release agent on mold	
Screw speed too high		
Sink Marks or Bubbles		
Injection pressure too low	Gate(s) too small	Flow too soft
Stock temperature too low	Mold temperature too low	Flow too stiff
Ram speed too slow	Sprue bushing too long	Excessively volatile content
Plunger dwell too short	Sprue diameter too small	Nonuniform external lubrication
Nozzle too cold	Insufficient venting	Excessive regrind mixed with virgin material
Nozzle bore too small	Runners too small	
Insufficient feed		

Reprinted by permission of Monsanto Company.

Figure 3-5. Six-head milk gallon blow molding machine. *(Courtesy of Uniloy)*

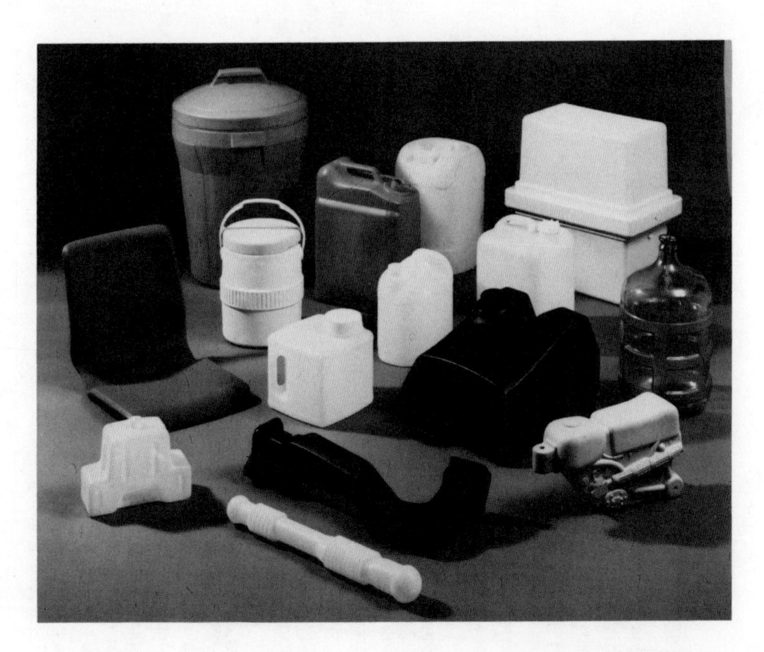

Figure 3-6. Miscellaneous blow molded articles. *(Courtesy of Uniloy)*

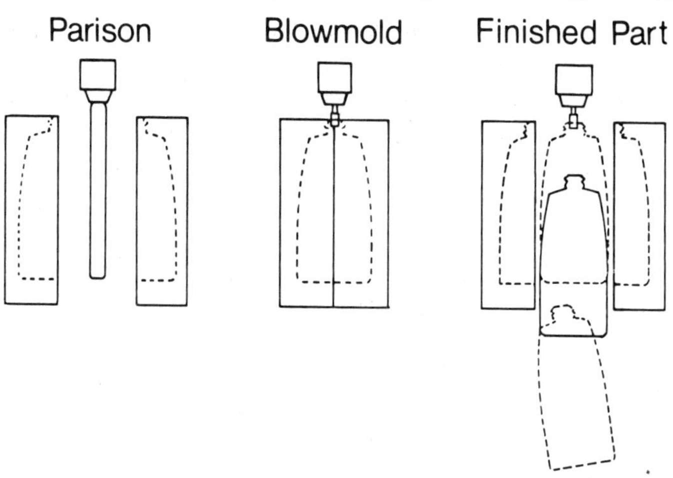

Figure 3-7. Blow molding sequence.

Extrusion–Blow Molding

In the extrusion–blow-molding process, the following sequence takes place in making a hollow part:

1. Forming of a parison (hollow tube) from molten plastic material
2. Entrapping the parison in the mold
3. Expanding the parison with air to form the part
4. Cooling the blown part
5. Removing the part from the mold

In many cases, this sequence can be completely automated, including the finishing, trimming, and removing the blown part.

Extruder. The extruder takes the material from the solid to the melt state. The extruder screw consists of three basic sections: a feed section, a transition section, and a metering section. Flights of the screw in the feed section are of constant diameter. The screw in this area simply is a conveyor that transports the plastic pellets to the transition or compression section. As the pellets move through the feed section, they absorb sufficient heat from the heater bands on the barrel wall nearly to reach the melting point. In the transition section, the root diameter of the screw gradually increases, reducing the area between the screw and surrounding barrel. The resulting compression melts the resin; the melt is then mixed and homogenized as it passes through the metering section.

There are many types of commercially available screws (single-channel, dual-channel, etc.) to satisfy the melting and mixing requirements of the different blow-moldable plastic materials.

Continuous Extrusion. Basically, the continuous extrusion process consists of a continuously rotating screw inside an extruder barrel. The rate of parison formation is synchronized by a variable speed control to the blow-molding cycle. Since the parison extrudes continuously, a separate blowing and cooling station is required in order to avoid interference of the clamp mechanism with parison formation. The parts are blow molded and cooled at the blowing station; the molds open, ejecting the parts, then return to the first station to engage a new parison (or parisons).

The movement of the molds can be reciprocating in a linear plane or reciprocating in a radial arc as long as the molds, when opened, have a clear path to return to the extrusion station and engage the parisons. Continuous parison extrusion provides a homogeneous melt with a uniform heat profile. Because there is no interruption in the parison formation, there is less chance for degradation of the melt as it passes from the extruder through the die head. This can be important when blow molding heat-sensitive materials. Multiple die heads can be used on continuous extrusion machines.

Molds

The molds perform the last major operation in blowing a plastic container. Each mold half is mounted on a moveable platen, which, when closed, centers the mold directly under the die and mandrel or blow pin.

The number of molds that can be used on a machine will depend on the size and type of clamp.

The mold has two basic functions to perform, namely, to form the part and to cool it. While the platens are open, the parison is extruded between the mold halves. When the parison reaches the proper length, the molds close and air pressure supplied through the die head assembly inflates the parison to the shape of the cavity. When the blown container has cooled, the air is exhausted and the platens open to allow the container to be ejected. Large industrial parts are often molded with a needle blow pin separate from the parison die head. The needle blow pin can be mounted on the side or bottom of the mold.

Most blow molds are either cast or machined aluminum with steel-pinch inserts.

Pinchoff is required at any point around the cavity where there is flash. This is a recessed area adjacent to the cavity and separated from it by a protruding edge. This edge compresses the flash from the container except for a very thin film and facilitates the easy trimming of the flash.*

Because the extruded parison is usually at a temperature of about 350°F, the molds are cooled. Normally this is done by passing chilled water and glycol mixture through channels drilled in the mold—top, body, and bottom sections. Optimum cooling of the blown part in the mold is important because it is the largest single time-consuming factor in the overall cycle.

The top and bottom cooling sections usually are separated from the body section since their cooling requirement is greater because of the pinchoff pockets.

Some of the latest technology for increasing the cooling rate in a mold include liquid CO_2, high-pressure moist air expansion, subzero air, and cycling air in and out of the blown part.

Finishing Containers

To form smaller parts and containers, automated processing can be used to deflash and finish the neck. Prefinish of container necks in the mold is the usual practice in container production. Neck, shoulder, and tail flash are removed in a secondary deflashing or finishing operation. An additional facing operation on the container lip also can be performed at this finishing station. The containers then can be conveyed directly to decorating and filling or packoff stations.

---

*Reprinted by permission of Uniloy.

Figure 3-8. Extrusion machine. *(Courtesy of Modern Plastic Machinery Corporation)*

EXTRUSION

Extrusion, like injection molding, is a process of melting and reshaping thermoplastic materials. Powders or pellets are fed through a hopper and into a heating chamber where the material is melted then forced through a die that shapes the mass into a continuous configuration. A screw, located in the plasticating chamber, melts, meters, and forces the melt through the die. Shapes such as rods, tubes, filaments, flat sheets, and profiles are extruded in a continuous operation. Another large outlet for extrusion is the coating of wire and cable for insulative protection. Other uses of extruders include the depositing of continuous film on substrates such as paper and foil and the supplying parisons for blow molding.

The most widely used extruder is the single screw, constant diameter machine. Two dimensions are used to express single-screw-extruder size: inside diameter of the barrel (D) and barrel L/D ratio. An extrusion machine is shown in Figure 3-8.

4 Acrylic Fabrication*

Working with acrylic is a pleasant and satisfying experience. Because of its outstanding clarity, gemlike luster, and weather resistance it is used extensively in the production of light fixtures, display fixtures and outdoor signs. Sheet stock can be purchased crystal clear or in one of many transparent or translucent colors and shades. Some windows are also made of clear acrylic but because it is softer than glass its use for windows is restricted to applications where scratch resistance is not a factor.

Acrylic sheets can be sawed and bonded into a large variety of attractive end products. The following recommendations represent techniques currently being used by many fabricators.

Table Saws

1. The circular saw (table saw) is best for straight cuts.
2. Use an 8–10-in. diameter, hollow-ground, high-speed metal-cutting blade having the following specifications.

Thickness of Sheet (in.)	Blade Thickness (in.)	Teeth per in.
1/16, 1/8, 3/16	1/16 or 3/32	6–8
1/4, 3/8, 1/2	3/32	4–6
5/8 and over	1/8	3–4

Hollow-ground blades give a smooth cut without chipping. The teeth should be ground with zero-degree rake on a straight line to the center.

3. The sheet must be held down firmly to prevent chattering and chipping.
4. Use a fast tool speed and a slow feed so that only a small amount

of material is removed with each revolution. Do not force the sheet into the blade. This can melt the acrylic.

5. When thick sections are being sawed, a continuous stream of compressed air or a 10% solution of water-soluble oil can be directed at the blade to prevent overheating. In general, the use of a coolant provides faster, smoother cuts.
6. When feeding the material, apply most pressure on the side between the fence and the blade.

Band Saws

1. Band saws are used for irregular and curved cuts. When straight cuts are made, it is advisable to use a guide bar (fence).
2. Use a metal-cutting 3/16-in.- or 1/4-in.-wide blade with slight set and 10–18 teeth per in. A 3/16-in. blade is preferred for cutting sharp contours. A 3/8-in. or 1/2-in. blade is normally used for straight cuts and thick sheets.

Thickness of Sheet (in.)	Minimum Blade Width (in.)	Teeth per in.
1/8	3/16	18
1/4	3/16	14
1/2	1/4	10
1	3/8	8

3. Fast speeds can be used for sections under 1/2 in. thick. Over 1/2 in., the speed should be slowed down to prevent overheating.

Saber Saws

1. Clamp a straight board across the sheet, near the edge of the work bench. This will serve as a guide and help to minimize vibration.
2. Use metal-cutting blades with approximately 24 teeth per in.

Drilling

1. Drilling can be done on a drill press or with a portable drill, but a drill press should be used whenever possible. It provides adjustable speed and better feed control.

2. The cutting edges on standard twist drills are too sharp for drilling acrylic. Small flats must be ground onto both cutting edges to prevent chipping. The flats should be approximately 1/32 inch wide.
3. Suggested drill speeds:

Diameter (in.)	Drill Speed (rpm)	Diameter (in.)	Drill Speed (rpm)
1/8	3500	3/8	1200
3/16	2500	1/2	900
1/4	1800	5/8	700

4. Holes larger than 1 in. in diameter can be cut with a hole-saw or fly cutter.
5. Reserve special drills for cutting acrylic. Do not use them for cutting metal.
6. Reduce the feed just before the drill point breaks through the bottom surface. This will prevent chipping.

Sanding and Polishing

1. Sanding produces a matte or satin finish. Polishing restores the original luster.
2. Scratches must be sanded out before polishing.
3. Most conventional-type power sanders can be used for acrylic. They should be used at somewhat slower speeds than those used for sanding wood.
4. Wet or dry belts and sandpaper are used with water for sanding acrylic.
5. Sanding should start with the finest grit that will remove the scratches, then followed with progressively finer grits down to 500 grit. Water should be used freely, and the sandpaper should be rinsed frequently to keep it from clogging.
6. Polishing can be accomplished by machine or by hand and by use of an oxygen-hydrogen flame.
7. A milled edge usually can be polished without prior sanding, but a saw cut must be sanded or run through a jointer or shaper or hand-scraped before it can be polished.
8. For edge polishing, an 8 to 14-in.-diameter wheel consisting of several 1/4-in.-thick sections built up to a width of 2 to 3 in. is very

effective. The wheels are made of bleached muslin and have only one row of sewing around the arbor hole.

9. Solidly stitched wheels with concentric rows of sewing should not be used. They are too hard and can burn the edges of the acrylic.
10. Buffing compounds, applied to the buffing wheel, are used to bring out an optimum finish.
11. A cleanup wheel made up of soft cotton or flannel sections can be used (without a buffing compound) to further enhance the surface of buffed edges.
12. Wheels for surface polishing can be from 6 to 12 in. in diameter, built up to a width of 1½ to 2 in. They are made of soft, bleached muslin for the initial polishing and of Domet flannel for the final finish. The wheels should have only one row of sewing around the arbor hole.
13. Fine scratches can be removed by hand polishing with a cleaning compound and a soft flannel pad.
14. Flame-edge polishing is accomplished by passing a hydrogen-oxygen flame directly across the surface at a speed of approximately 3 to 4 in. per sec. A welding torch with a No. 4 or No. 5 tip is used. The hydrogen pressure is set at 8 lb and the oxygen pressure at 5 lb. The hydrogen is ignited first, then the oxygen. The flame should be bluish, nearly invisible, and approximately 3 in. long.

Chips and Cracks

1. Chipped edges can lead to cracks in the sheet. They should be removed by filing, beveling, scraping or sanding. Chipped holes should be countersunk.
2. If a crack has started at the edge of a sheet, a small hole located 1/16 in. away will prevent propagation of the crack.

Cleaning

1. Acrylics must never be cleaned with solvents such as acetone and MEKP because these solvents attack the surface. The protective masking paper should be left on as long as possible. All layout work can be done on the paper surface. Soap or detergent and water usually wash away most contaminants. They should be

applied with a soft cloth and light pressure then dried with a clean, damp chamois.

2. Cleaning a dirty acrylic surface with a dry cloth will scratch the surface.
3. Hexane or naptha are sometimes used to clean acrylic in industry.
4. Acrylic sheets containing masking paper should never be left in the sun for long periods. A few days' exposure will harden the adhesive, making it almost impossible to remove.

Solvent Cementing

1. Acrylic sections can be cemented together to form transparent joints and strong assemblies.
2. Bonding edges with a smooth saw cut do not normally require further preparation. Those with irregular surfaces must be milled, wet sanded, or finished on a jointer or shaper to provide a flush fit.
3. Edges to be cemented should not be polished. Polishing produces convex-shape edges that destroy the necessary surface flatness.
4. Cementing should not be attempted at temperatures below 65°F.
5. In solvent cementing, one edge is soaked in a solvent such as methyl chloride or ethylene dichloride for 1–3 min, then placed and held flush on the other acrylic adherend for 1–3 min. The parts are held together without pressure for approximately 30 sec to allow the solvent on the soaked edge to attack the other surface. After 30 sec, gentle pressure is applied to force out all air bubbles and effect a transparent joint.
6. Allow cemented joints to remain under pressure for at least 30 min.
7. Small pieces of acrylic can be dissolved in solvent to form a viscous cement. The general procedure for bonding with viscous cements is similar to methods for bonding with solvents.
8. Parts that are too large for available trays can be solvent-soaked on a glass plate or sheet of galvanized steel. Several brads or wires are placed under the edge, and the solvent is fed in from both sides with a medicine dropper.
9. Formed letters should be sanded before bonding because the back edges are usually uneven.

---

*Reprinted by permission of CY/RO Industries.

5 Cellulosic Fabrication

FABRICATION OF CELLULOSICS (ACETATE, BUTYRATE, AND PROPIONATE)

Machining

Cellulosics can be worked with most tools used for machining wood or metal. Tool speeds should be such that the plastic does not melt from frictional heat. The highest speed short of overheating will give the best results. It is also important to keep cutting tools sharp at all times; and hard, wear-resistant tools with greater cutting clearances than metal-cutting tools are suggested. High-speed or carbon-tipped tools are efficient and economical, especially for long runs.

Because plastics are poor heat conductors, the heat generated by machining operations must be absorbed by the tool or carried away by a coolant. A jet of air directed at the cutting edge aids in cooling the tool and also serves to remove chips. Oil, soapy water, and plain water are sometimes used for cooling, and they also promote a smooth cut. These coolants cannot be used, however, if trim scrap is to be reused. Trim scrap contaminated with oil or other foreign matter cannot be molded satisfactorily.

Drilling

Drills designed especially for plastics are suggested for drilling cellulosic plastics; but standard twist drills as used for drilling metal or wood are often satisfactory.

Wide, highly polished flutes are desirable because they expel the chips with less friction and therefore tend to avoid overheating and consequent gumming. Drills with substantial clearance on the cutting edge of the flute make a smoother hole than those with less clearance. Drills should be backed out often to free chips, especially when drilling deep holes. Peripheral speed of twist drills for plastics ranges from 100 to

200 ft (30.5 to 61 m) per min. The rate of drill feed into the plastic varies from 0.010 to 0.025 in. (0.254 to 0.64 mm) per revolution. The flow of the plastic governs the drill speed and feed rate that should be used. High speeds and high feed rates are used for the soft-flow materials and low speeds and low feed rates for the hard-flow materials.

Reaming. Reaming holes drilled in articles made of acetate, butyrate, and propionate is not recommended. Where close tolerances are required in thin sections, good results can be obtained by drilling to within 0.001 in. (0.025 mm) of desired size and then pushing a hardened, polished rod through the hole to obtain a smooth surface.

Tapping Internal Threads. Conventional four-flute taps can be used for cutting internal threads when close fits are required. Such taps, however, have a tendency to generate considerable heat during the tapping operation.

A high-speed steel, two-flute tap is suggested for longer life and greater tapping speed than conventional taps. The two-flute tap provides greater clearance for chip discharge. Flutes should be ground so that both cutting edges cut simultaneously; otherwise the thread will be ragged and rough. Cutting edges should be 85 degrees from the center line, giving a negative rake of 5 degrees on the front face of lands, so that the tap will not bind in the hole when it is backed out. Some relief on the sides of threads is desirable. Taps should have a two-thread-taper cutting lead. Either the tap or the work should be free to center.

Lathe Operations

Cellulosic plastics can be readily turned or threaded on a lathe. In general, turning speed should be as high as practicable because cuttings tend to form long threads that coil around the work at low peripheral speeds. At high peripheral speeds, centrifugal force causes the thread cuttings to be thrown away from the work. The maximum satisfactory lathe speed is ordinarily about 700 rpm.

Milling Operations

Cellulosic plastics can be machined satisfactorily with standard high-speed milling cutters as used for metal, if the cutters have sharp edges and adequate clearance at the heel. Tolerances can be easily held to 0.002 in. (0.051 mm) with maximum cutting speeds.

Sawing

Cellulosic plastics can be sawed with any of the saws used for wood or metal—circular saws, band saws, saber saws, jigsaws, hacksaws, or handsaws. Some of these tools, however, are better suited than the others for sawing plastics because they produce smoother or faster cuts than others. Circular saws and band saws usually produce the best surfaces, and they can be used in most sawing operations.

Blade design plays an important part in the successful sawing of plastics. Band-saw blades must have set teeth for any type of sawing, but for curved cuts the blade should be narrower and have more set than a blade used for straight cuts. The blade should be just soft enough to permit filing, and it must be kept sharp to prevent melting or chipping of the plastic. The blade guide should be placed as close as possible to the material being cut.

For straight cuts, a circular saw, even though it tends to generate more heat than a band saw, is preferred because it produces a smoother cut. A perforated saw blade will run cooler than a solid blade. It is essential that the spindle bearing of a circular saw be tight so that the saw runs true. The saw should be hollow-ground with no set to the teeth. For best results, the tooth pitch should be to the center of the arbor hole with a well-rounded gullet to permit free curling and ejection of chips.

A clean cut with minimum breakage or chipping after cutting through the plastic may be obtained by using a cross-cut circular saw operated at maximum speed. The saw should have radial teeth in sets of four that are ground as follows: the first, a straight-ground chisel tooth with edge parallel to the axis; the second, a tooth ground with the right point high; the third, another straight tooth; the fourth, a tooth ground with the left point high. The straight teeth, or teeth 1 and 3, should be 0.001 in. (0.025 mm) lower than the taper-ground teeth.

Shearing, Blanking, and Punching

Sheets of acetate, butyrate, and propionate can be sheared, blanked, or punched with sharp-edge cutting tools. In any of these operations, warming the sheet is essential if a smoothly cut edge is to be obtained. The greater the thickness of the material, the higher the warming temperature required. However, the material should not be heated to the point that its lustrous surface is damaged. When a sheet of plastic is heated, it softens, and the pressure required to shear, blank, or punch the sheet slightly compresses the material. After the pressure is released from the warm material, the sheet expands to its original size, leaving one edge slightly concave and the other convex. In shearing, blanking, or punching, it is virtually impossible to obtain both a smooth cut and a straight edge at the same time, particularly on a thick sheet. The curved edges are scarcely noticeable on sheet less than 0.100 in. (2.54 mm) thick.

Sanding

Cellulosic plastics may be sanded in production operations on machines made for this purpose. These machines, of both bench and upright types, have an endless abrasive belt that runs at a speed of approximately 2,000 ft (610 m) per min. Both coarse and fine abrasive belts are used. Frequently, preliminary sanding is done on a coarse belt, and final sanding on a fine one. In order to avoid gumming of the material through overheating, the work should be pressed only lightly against the belt. It is often desirable to sand one piece partially, then go to the next, and so on, coming back to the first piece after it has cooled.

Tumbling

Flash is seldom removed from cellulosic plastics by tumbling. Cellulosic plastics are so tough that if they are tumbled to remove flash, the surface of the entire article is abraded and must be polished. However, the addition of dry ice to the tumbling barrel causes flash to become relatively brittle and, in some cases, a short period of tumbling can be used to remove thin flash without serious impairment of the surface finish of the molded article.

Solvent Polishing

A very effective method of producing a high gloss and eliminating fine surface scratches on cellulosics is solvent polishing. The article is usually dipped into a suitable solvent and then suspended to dry. The dipping time should be as short as possible, the surface usually requiring only enough solvent to wet it with an even film. If a quick-drying solvent is used, it is often necessary to add a small amount of slow-drying component to prevent humidity-blush after drying.

Excessive surface strains or internal strains may cause the surface of an article to become rough when dipped into an active solvent. This type of rough surface, commonly referred to as *orange peel,* is more prevalent on injection-molded items than extruded or compression-molded items. An article to be solvent polished should be spot-checked on a hidden area to determine whether enough strains are present to cause orange peel.

Some solvent mixes that have been used successfully for dip polishing are as follows:

Acetate	Butyrate and Propionate
70% acetone	70% isopropyl alcohol
30% ethyl lactate	30% toluene

The specific gravity of the solutions containing acetone should be determined with a hydrometer and maintained as nearly constant as possible by adding acetone from time to time. Acetone has a high evaporation rate and will be lost from the solution over long periods of time.

Another effective method of polishing cellulosic plastics is to wipe them with a cloth wet with one of the above-mentioned solutions. The cloth should be resaturated with the solution before wiping each article. This method has the advantage of a very short drying time.

Some molders use a solvent-vapor method of polishing molded articles. This works by vaporizing a solution, such as 69% ethyl acetate and 31% ethyl alcohol, and passing the molded article through the vapors. Of course, adequate ventilation should be provided to prevent excessive inhalation of vapors by personnel in the area.

Cementing

In the cementing of acetates it is necessary that several precautions be observed.

1. The surfaces to be joined must be clean. A slight film of oil, water, or polishing compound may cause poor bonding.
2. The surfaces must be smooth and as well aligned as practicable.
3. The solvent or cement must be sufficiently active to soften the sur-

face to a depth that will allow a small amount of flow to occur when pressure is applied to the softened area.

4. The solvent or cement must be of such composition that it will dry completely without blushing.
5. Pressure must be applied until the cemented joint has set to the extent that no movement occurs when the pressure is released.
6. Subsequent finishing operations must be delayed until substantially all of the residual solvent has dissipated.

Two types of cementing agents are in wide use: those consisting of solvents only (solvent type), and those consisting of a cellulose ester or other polymer dissolved in a solvent blend (dope type). In general, when the surfaces to be cemented are in a single plane, the solvent type is used. When the surfaces are irregular or located so that applying the solvent is difficult, the dope type is used.

Cementing Technique with Solvents. Small articles with plane surfaces to be joined can be cemented satisfactorily by holding the articles together with one hand and applying the solvent along the edges of the joint with a hypodermic needle, medicine dropper, or small oil can, and making sure that the solvent flows throughout the area to be cemented. The joined articles can be safely laid aside to dry soon after the solvent has been applied.

For large articles, the preferred method of cementing with a solvent is to immerse the surfaces to be joined in the solvent until the material is softened, and then to clamp and hold them in position until the bond has set. A convenient method of application is to hold the article in such a position that the surfaces to be cemented barely touch the solvent. This may be done by maintaining a constant level of solvent in a shallow pan and supporting the article on a pad of polyurethane foam or felt. The pad acts as a wick, thus allowing light contact of the article with the solvent.

If polyurethane foam is used, it is suggested that the pad consist of two layers, each about ¼ in. (6.4 mm) thick. If a felt pad is used, a fine wire screen over the pad helps to minimize contamination of the plastic surface with fibers from the felt. To reduce the rate of evaporation, the pad may be covered with a closed tray from which the top has been cut in so that one or more of the articles to be joined will fit in the openings. When an article is removed for joining, another is put in its place, thereby keeping the pad fully enclosed most of the time.

In order to secure a bond as strong as the plastic articles themselves, it is necessary to soften substantially the edges to be joined and then clamp the articles together. Although this procedure may cause the softened plastic to protrude at the joint, the protrusion can be removed (after the pieces have stood 24 to 48 hours to permit the residual solvent to dissipate) by suitable machining operations followed by polishing. If a high finish is desired, it may be necessary to ash the surface at the joint after it has been machined and before it is polished.

Solvents. Acetone is one of the most commonly used solvents for cementing cellulose-ester plastics. However, it is ordinarily not good practice to use acetone or other low-boiling solvents alone because they evaporate so quickly. Rapid evaporation of solvent is likely to cause moisture blush—a white frosty appearance of the cemented joint. Also, low-boiling solvents may evaporate before they have had sufficient time to soften the surfaces and effect adhesion. Blushing is avoided and evaporation reduced by adding to the low-boiling solvent one or more solvents of higher boiling points, of which the following solvent mixtures are common examples.

Acetate	Butyrate and Propionate
1. 70% acetone	80% butyl acetate
30% ethyl lactate	20% butyl lactate
2. 30% ethyl acetate	—
40% acetone	—
30% ethyl lactate	—

Cementing Technique with Dopes. The preferred technique for cementing with dopes is to prepare the surfaces carefully, as previously discussed, and then apply the cement with a brush or other mechanical applicator. It is frequently desirable to soften the surfaces to be joined by an application of undiluted solvent before applying the dope cement. Subsequent handling is the same as when using solvents alone, and dopes can also be used in the same manner as solvent blends, provided the viscosity is not too high. Suitable cements for acetate, butyrate, and propionate may be obtained by making a 10% solution of the plastic in one of the appropriate solvent mixtures described above.*

*Reprinted by permission of Eastman Chemical Products, Inc.

6 Miscellaneous Thermoplastic Processes

PLASTISOL MOLDING

As noted in chapter 2, plastisols are a mixture of fine vinyl particles and plasticizers. They are used in a variety of dipping and coating processes. The basic principle of plastisol fabrication is the fusing together of the plastisol particles to form a tough, flexible coating on a male mandrel. The cured coating is either left on in a protective coating or removed as a functional article. Wire racks and tool handles are among items coated with plastisols by the dip-molding process. The coating formation occurs when the item is heated to approximately 325°F then dipped in the plastisol. The vinyl particles fuse together to form a smooth, continuous coating. Coating thickness is dictated by the temperature of the immersed article: the higher the temperature, the thicker the coating. Withdrawal after dipping is performed at a controlled rate to insure smoothness and uniform thickness. Parts are usually ejected from molds by compressed air. Among products fabricated by plastisol molding are rainboots, gloves, closures, toys, and spark-plug covers.

HEAT SEALING

The basic principle of heat sealing involves the melting of two or more thermoplastic films and fusing them by the application of pressure. The films are pressed between two narrow contact surfaces, one of which is electrically heated. The apparatus remains in the closed position for approximately 10 seconds after the seal has been made to allow the joint to cool down. Premature opening results in a broken or incomplete seal.

Some sealers have automatic movement, others are actuated by a foot pedal; and in some instances small hand irons are used.

Heat sealers are employed extensively in packaging foods, textiles, toys, and hardware. Pedal-actuated models are used in the aircraft and aerospace industries for sealing vacuum bags.

In many of the packaging applications, thermoplastic films are sealed to cardboard: the molten film is imbedded into the cardboard and a strong bond forms between the two materials.

CALENDERING

Calendering is the process of squeezing a softened thermoplastic between two or more rolls to form a continuous film. After the heated mass is fed between heated rolls that are preset for desired thickness, the film goes through cooling rolls and then on to the wrapping roll. Polyvinyl chloride is used most often for calendering operations.

WELDING PLASTICS

There are several methods for fusing and welding plastics including friction welding, hot-gas welding, fusion welding, and ultrasonic welding. They all, more or less, have the same requirement: the materials used must be thermoplastic, and mating surfaces must be sufficiently heated so they can be fused together.

ENGRAVING PLASTICS

This process is used extensively in the production of nameplates and signs. Laminated thermoplastic sheets having two or more layers of different colors are used. The tool of the duplicating machine, called a *pantograph machine,* cuts through the top layer, showing the design or lettering in color of the second layer. Most pantograph machines are equipped with a variable reducing feature that permits the engraving work to be smaller than the model while retaining the same proportions. Pantograph machines are also used to cut brass stencils for marking part numbers, serial numbers, and the like on components and assemblies.

CARVING PLASTICS

This process involves the cutting of a design in an acrylic block or sheet for decorative purposes. After the design has been completed, it is treated with a dye to achieve the desired color. The combination of acrylic clarity with a colorful floral design makes a very attractive display. Cutting is done either by a pointed vibrating tool moving up and down or by a rotary tool with a pointed twist drill. For added attraction, a colored layer can be bonded to the carved surface. Carving is not only an interesting hobby, but can also be profitable. People who are adept at carving plastic can make a design in a small acrylic plate in a matter of seconds.

ROTATIONAL MOLDING

Rotational molding is the process in which thermoplastic powders are loaded into closed, hollow molds and rotated biaxially in a heated chamber. The heat melts the powder so that a uniform casting develops on the cavity walls. The assembly, still rotating, is then transferred to a cooling chamber where the temperature is lowered to a point allowing the liquid to become a solid that can be handled. Here again, the shape of the mold dictates the shape of the part. Average cure cycles usually range from 6 to 10 min.

EDGE LIGHTING

Edge lighting is the phenomenon whereby light enters the sanded end of an acrylic rod or sheet, travels between highly polished surfaces, and emits through another sanded area, the location of which is dictated by the application. For throat lights, one end of a highly polished acrylic rod is recessed and threaded to accept a small flashlight designed specifically for this application. The light enters the sanded surface and comes out through the opposite end, which is also sanded. This type of light is very useful for penetrating visually blocked areas. Light can also travel through curved acrylic, provided the curve is not too severe.

A polished acrylic plate with a carved and colored facing can be mounted in a lighted box to form an attractive display. The end entering the light box is sanded and the opening in the box is just large enough to accommodate the plate. This type of display is especially attractive in a darkened room. Large edge-lighting displays are occasionally used in nightclubs.

HOT STAKING

Similar to riveting with metals, hot staking is a process used to fasten plastics to plastics mechanically. A thermoplastic rod inserted through holes in the materials to be joined is melted at each end with a hot tool,

Figure 6-1. Continuous high-speed, roll-fed thermoforming machine. *(Courtesy of Brown Leesona Corporation)*

Figure 6-2. "Pellets to Parts" thermoforming machine. Combines extruder, thermoformer, printer, grinder, trim press and blender all in one compact system. *(Courtesy of Brown Leesona Corporation)*

leaving a rounded, flanged head at the point of contact. With molded products, the stakes are usually molded into the part.

THERMOFORMING

Thermoforming involves heating of a thermoplastic sheet beyond its distortion point then forcing it over a preshaped mold so that when cooled it retains the shape of the mold. Pressure on the sheet is usually applied by vacuum, but can also be applied by air or mechanical means. Figure 6-1 shows a continuous high-speed thermoforming machine. Figure 6-2 shows an extruder-type thermoformer.

7 Curing Systems for Epoxies and Polyesters

EPOXY RESINS AND CURING AGENTS

Most curing agents used with epoxy resins are 100% reactive so that they actually become part of the newly formed compound. Diluents such as phenyl glycidyl ether and butyl glycidyl ether are used to thin down thicker epoxy resins. They too are reactive.

The cure shrinkage of epoxies is low because they do not give off any by-products during cure. This also helps to account for their excellent dimensional stability. Cured epoxies exhibit low water absorption and good resistance to acids, solvents, and alkalis.

Table 7-1 lists the leading epoxy resin manufacturers in the country. It also shows their commerical designations. Some of the most common curing agents used with these resins can be found in Table 7-2.

Some of the characteristics of cured epoxies such as heat and chemical resistance are derived from the curing agent. When one of the resins in Table 7-2 is cured with DTA, the heat distortion point of the cured part is approximately 225°F. The same resin cured with metaphenylene diamine will produce a part with a heat distortion point in excess of 300°F. Two epoxy resin systems that do increase heat resistance are the novolacs and the cycloaliphatics.

Care should be exercised when working with epoxy curing agents, especially the room-temperature-curing amines. Prolonged contact of the skin with these agents can cause dermatitis. Some of the low-viscosity resins can also be a problem. As a safeguard, rubber gloves or an appropriate skin cream should be used when working with these materials.

POLYESTER RESINS AND CURING AGENTS

Polyester resins are used extensively in molding compounds and as binders for fiberglass-reinforced plastics (FRP). They can be cured at room

Table 7-1. Epoxy Resins

Manufacturer	Viscosity 11-16000 CPS	Viscosity 600 CPS
Shell	Epon 828	Epon 815
Ciba-Geigy	Araldite 6010	Araldite 506
Union Carbide	ERL 2774	ERL 2795
Dow	DER 331	DER 334
Reichold	EPI-TUF 37-140	EPI-TUF 37-130
Celanese	EPI-REZ 510	EPI-REZ 5071

temperature by the addition of a catalyst and an accelerator or at elevated temperatures by the addition of a catalyst and heat.

Curing Systems

Methyl ethyl ketone peroxide (MEKP), with addition of a promoter such as cobalt naphthenate, is the most commonly used catalytic system for room-temperature cures, while benzoyl peroxide (BPO), used without an accelerator, is most popular for elevated-temperature cures. MEKP is a thin, water-white liquid. BPO is a solid but is mixed with liquids such as tricresyl phosphate to form a white paste. Cobalt naphthenate is a purple liquid containing 6% cobalt.

Benzoyl peroxide is normally used in the proportion of 2% of the weight of resin for vacuum-bag lamination. The pot life is approximately 4 hours. This allows ample time for the layup and bagging operations. If a longer pot life is required, a catalyst such as dicumyl peroxide can be used. The cure temperature, however, will be higher. For room-temperature-curing systems, see Table 7-3.

Working with Cobalt. Because of the explosive nature of a direct peroxide-cobalt mix, cobalt naphthenate is often added to polyester laminating resins by the manufacturer. Practically all casting resins sold in retail stores contain cobalt naphthenate. This is to protect the hobbyist engaged in decorative embedding. Care must still be exercised when working with this combination because, if a large disproportional amount of MEKP is added to cobalt-promoted resin, the mix could get hot enough to smoke and even catch fire. Cobalt gives resin a slight purplish cast. For clear-casting resins, the purple is neutralized.

Table 7-2. Epoxy Curing Agents

Curing Agent	Liquid or Solid	Concentration (phr)*	Cure Time	Cure Temperature (F°)	Pot Life	Approximate Distortion (F°)	Comments
Diethylene triamine (DTA)	L	8–12	—	R.T.	30 min	230	Quick curing at R.T.
Triethylene tetramine (TETA)	L	10–14	—	R.T.	30 min	230	Quick curing at R.T.
Diethylaminopropylamine (A) (DEAPA)	L	4–8	30 min	240	3–4 hr	230	Longer pot life than DTA
Tridimethyl aminoethyl phenol (DMP-30)	L	6–10	—	R.T.	40 min	220	Quick curing used with anhydrides as accelerator
Curing agent T (Shell)	L	15–25	—	R.T.	15–20 min	200	Considerably lower level of toxicity than DTA or TETA
Curing agent Z (Shell)	L	20	1 hr	210	4–6 hr	295	Excellent chemical resistance, good heat resistance
Piperidene	L	5–7	6–12 hrs	300	6–10 hr	230	General-purpose curing agent
Methyl nadic anhydride (MNA)	L	60–80	2 hr	400	1–3 hr	250	Requires accelerator
Metaphenylenediamine (MPDA)	S	14	1 hr	210	4–6 hr	315	Excellent chemical resistance, good heat resistance
Diaminodiphenyl sulfone (DDS)	S	30	1 hr	300	1–3 hr	345	Good heat resistance
Dicyandiamide (DICY)	S	4	30 min	345	1 yr	230	Long shelf life used with solid epoxies
BF3-400	S	2–4	1 hr	250	1 mo	340	Used as accelerator for various anhydrides (IPHR)

Phthalic anhydride (PA)	S	40–50	8 hr	300	2–6 hr	285	Low exotherm, used with an accelerator
Hexa hydrophthalic anhydride (HHPA)	S	75–85	2 hr	225	6 days	250	Good impact strength
Pyromellitic dyanhydride (PMDA)	S	To 55	8 hr	425	extensive	510	High temperature resistance, does not require accelerator
Dodecenyl succinic anhydride (DDSA)	L	120–150	6 hr	200	3–4 days	240	Semi-rigid, requires accelerator
Benzophenonetetracarboxylic dyanhydride (BTDA)	S	60–80	2 hr	390	—	545	High temperature resistance
General Mills Polyamides							
Versamid 115	L	50R TO 50V	24 hr	R.T.	50 min	140	No toxicity, low exotherm, concentrations
Versamid 125	L	60R TO 40V	24 hr	R.T.	50 min	140	Can be increased if semi-rigid pottings are desirable
Versamid 140	L	70R TO 30V	24 hr	R.T.	50 min	140	

*Parts per hundred resin.

Table 7-3. Catalysts and Promoters for Polyesters

Room Temperature Curing Systems Catalyst	Promotor
Methyl ethyl ketone peroxide (MEKP)	Cobalt naphthenate *or* cobalt octoate
Cyclohexanone peroxide	Cobalt nephthenate *or* cobalt octoate
Cumene hydroperoxide	Manganese naphthenate
Benzoyl Peroxide (BPO)	Dimethyl aniline

Recommended concentration

Type	Curing System (phr)*	Concentration	Gel Time (mins)	Time to Reach Barcol 35 (hr)
Gel coat (filled)	MEKP	1.5	30–45	Layup is applied after 40 min
	Cobalt naphthenate	0.5		
Standard Layup	MEKP	1.0	25–35	6–8
	Cobalt naphthenate	0.5		
Fast cure layup	MEKP	1.0	15	2–3
	Cobalt naphthenate	0.5		
	Dimethyl aniline	0.1		

*Parts per hundred resin.

Diluents and Inhibitors. Styrene is usually added to polyester-laminating resins to lower the viscosity of the resin. Styrene is actually a monomer and becomes part of the cured resin structure. Thinner resin permits more thorough wetting of laminate fibers and assists in removal of trapped air. Diluents also reduce the cost of low-pressure molding compounds by allowing greater use of fillers. Also added to some polyester resins are inhibitors such as hydroquinone that prevent premature gelation of the resin.

Surface Tack. Cast polyesters usually have a slightly tacky surface after cure unless covered by cellophane or other appropriate film. For multiple layers, a few drops of a solution of styrene and wax to the top pouring will eliminate the tack. The wax rises to the top and forms a barrier that prevents air from contacting the surface of the resin. The surface, of course, will have a grainy appearance, due to the presence of the wax. Wax surfacing is usually used when the top pour contains an opaque coloring. The first pour in the mold will be to the top of the casting. The last pour will be the bottom. In the absence of a wax surfacer, tack can be removed by heating the casting in an oven at a low temperature (120–150°F) for several hours.

8 Reinforcements for Thermosets

REINFORCEMENTS FOR THERMOSETS

Though many thermoplastics are being reinforced today, reinforced thermosets still dominate in high-strength and high-temperature-resistant applications.

The use of short fibers and powdered fillers in molding compounds adds a degree of strength and stability to molded parts, but the strongest reinforced plastics are those made by filament winding and lamination.

Laminate Composition

Most laminates are made with woven cloths or some form of fiberglass mat. High-strength, unidirectional tapes such as graphite fiber tapes are also finding increased use. With these tapes, the fibers for each ply of buildup are oriented in a predetermined direction for maximum strength and stability. The reinforcement in a laminate is known as the *carrier.* The catalyzed resin used to impregnate and bond together the carrier plies is called the *binder.* Woven fabrics are designed to give the most strength where needed through control of the ratio of *warp* strands to *fill* strands. Warp strands are those running in the roll direction; fill are those in the width direction.

Most carriers are fibrous reinforcements composed of materials such as fiberglass, high silica, graphite, asbestos, boron, carbon, high-strength thermoplastics, and quartz.

Fiberglass is the most widely used reinforcement for plastics. There are several grades of fiberglass, including E glass (electrical), C glass (chemical), and S glass (high-strength). E glass is the all-purpose fiber and most commonly used. S glass has a tensile strength of approximately 700,000 psi as compared to approximately 500,000 psi for both E and C glass. When E glass is acid leached and re-fused at high temperatures, it becomes much more temperature resistant, but also suffers a drop-off in strength. The newly formed material is known as *high silica* and is used in high-temperature and ablative applications.

Fiberglass is used extensively for products such as boats, car bodies, trays, housings, construction materials, plastic tooling, airplane, and rocket parts.

Graphite, carbon, and *boron* fibers provide high tensile strength and modulus. Graphite, in particular, is being used in increasing volumes by the aircraft and aerospace industries. It is lighter and stiffer than fiberglass and can withstand much higher temperatures. It is indeed the most promising engineering reinforcement for the future.

Kevlar 49,* a lightweight, high tensile strength polyamide (nylon) developed by DuPont is also attracting much attention as a high-performance reinforcement for laminates and belted tires. It is more resistant to abrasion than are other high strength fibers. Kevlar fibers are also known as Aramid.

Ablative reinforcements, in addition to high silica, include graphite, quartz, fiberglass, and carbon fibers. Ablative resins include phenolics, silicones, and phenylsilanes. Ablative reinforcements improve strength, erosion resistance, and char formation.

Reinforcements for Filament Winding

With few exceptions, most filament-wound products have been made with fiberglass. The high tensile strength, comparatively low cost and ease of handling of fiberglass rovings have led to its virtually unchallenged monopoly in this area. The single weakness of fiberglass is its inherent low modulus. With reduction in cost, however, the currently more expensive, exotic, high-modulus fibers may eventually present a challenge to the supremacy of fiberglass. Discovery of new high-strength, low-cost fibers is also a probability.

FIBERGLASS TECHNOLOGY

High Strength

The contribution of high tensile strength of fiberglass fabrics to composite strengths (200,000–220,000 psi) means that fiberglass fabrics

*DuPont trademark.

have higher strengths than many other man-made textile fabrics. The breaking strength of fiberglass fabrics ranges from a few pounds per inch to several hundred pounds per inch of fabric width, depending on construction and yarn content.

Heat and Fire Resistance

Being inorganic, fiberglass fabrics will neither burn nor support combustion. This results in their use without deterioration at elevated temperatures.

Chemical and Weather Resistance

Fiberglass fabrics are resistant to attack by most chemical solutions and extreme weather conditions, and are almost completely unaffected by fungus, bacteria, and insects.

Moisture Resistance

Fiberglass fabrics do not absorb moisture; therefore, they are dimensionally stable, resist rot, and retain maximum strengths in humid environments.

Thermal Conductivity

The thermal properties of fiberglass fabrics prevent hot spots and permit the maintenance of temperature control in thermal environments.

Electrical Properties

Being nonconductive, fiberglass fabrics are ideal materials for use in electrical insulation. The thermal stability of fiberglass fabrics at elevated temperatures makes these fabrics ideal as reinforcements for insulating varnishes for use at high operating temperatures.

Dimensional Stability

Because the maximum elongation of fiberglass is very small, fiberglass fabrics neither stretch nor shrink.

Making Glass Fibers

The ingredients in the glass formulation are dry-mixed and melted in a high refractory furnace at temperatures of 2,300°F (1,260°C). The molten glass is gravity-fed through holes in the base of a platinum-alloy tank or bushing, gathered together, and drawn or stretched mechanically to the proper dimensions.

To prevent damage to the filaments during subsequent manufacturing processes, which can include twisting, plying, and weaving, a binder composed of starch and lubricating oils is applied to the filaments as they leave the bushing.

Glass Composition

Fiberglass yarn, like any other form of glass, is composed of silicon dioxide and various other oxides that determine specific properties. Proportions of these fiberglass ingredients in E glass and S glass are as follows:

Fiberglass Composition	E Glass (electrical) (%)	S Glass (high-strength) (%)
Silicon dioxide	54.3	64.2
Aluminum oxide	15.2	24.8
Ferrous oxide	—	0.21
Calcium oxide	17.3	0.01
Magnesium oxide	4.7	10.27
Sodium oxide	0.6	0.27
Boron oxide	8.0	0.01

GLASS FABRIC CONSTRUCTION AND WEAVES

The properties of fiberglass fabrics depend primarily on the fabric construction consisting of fabric count, warp, and filling yarn content and weave pattern. These factors determine fabric characteristics such as weight, thickness, and breaking strength.

Fabric Count

Fabric count is defined in the glass system as the number of warp yarns or "ends" per in. of fabric width running lengthwise in the fabric, and the number of filling yarns or "picks" per in. of fabric width running crosswise in the fabric.

In the metric system, fabric count identifies the number of ends or picks per 5 cm, thus yielding whole number counts when determined on the 5-centimeter specimen width.

Weave

The weave pattern describes the manner in which the warp yarns and the filling yarns are interlaced in the fabric.

Plain weave consists of one warp end woven over and under one filling pick. Plain weave is generally characterized by fabric stability with minimum pliability, except at low fabric counts.

Basket weave is similar to the plain weave, but has two or more parallel warp ends weaving together over and under two or more filling picks weaving together. The basket weave is more pliable, flatter, and stronger than the plain weave, but tends to be less stable.

Leno weave is produced by interlocking two or more parallel warp ends over each other and interlocking with one or more filling yarns. It is useful in reducing sleaziness in low-count, openly woven fabrics.

Twill weave is constructed with one or more warp ends weaving over and under two, three, or more filling picks in a regular pattern. This produces either a straight or diagonal line in the fabric.

Crowfoot satin weaves are constructed with one warp end weaving over three and under one filling pick. These weaves are more pliable than either plain or basket weaves, having conformability to complex or compound curved surfaces, and making possible the weaving of higher counts than plain or basket weaves.

Long shaft satin weaves have one end weaving over four or more and under one filling pick. Satin weave fabrics are the most pliable and conform readily to compound curves, produce laminates and reinforced moldings with high strength in all directions, are woven in the highest construction or density, and are less open than other weaves.

Finishes and Finishing

Untreated, or loomstate, fiberglass fabrics contain the binder originally applied during yarn manufacture. The presence of this binder is satisfactory for some applications but is not suitable for others. Therefore, binders must be removed before treatment with other chemicals formulated to render the fabric useful for industrial products such as composites. These yarn binders are removed under carefully controlled time-temperature conditions either by a batch heat cleaning or by a continuous *coronizing* process.

Chemical coupling agents are applied to heat-cleaned fiberglass fabrics to provide interfacial bonding between resins and fiberglass surfaces. These finishes also speed resin penetration into the fiber bundles, thus affording excellent resin "wettability."

Finishes for other applications can be applied to either untreated or heat-cleaned fabric. Weavesets with various thermoplastic resins, elastomers, or rubbers act as tie coats for coatings. Thermally resistant lubricating finishes are applied to heat-cleaned fabrics for high-temperature filtration where particulate free exhaust gases are demanded.

APPLICATIONS*

Electrical

The intrinsic electrical and thermal properties of fiberglass fabrics were instrumental in their early acceptance in high-temperature electrical insulation. Combined with alkyd varnishes, silicone resins, silicone rubber, PTFE, mica, or pressure-sensitive adhesives, and slit into tapes, these products continue to be used widely in electrical equipment of all kinds.

Fiberglass fabrics in combination with thermosetting epoxy resins constitute the basic raw material for printed circuit boards, where the thermal, electrical, dimensional stability, and mechanical property contributions of fiberglass fabric to the laminate find maximum utilization. Lightweight fiberglass fabrics are employed extensively in multilayer printed circuitry, imparting a high degree of dimensional stability to products of this kind as well as to flexible electrical products.

Reinforced Plastics

Strength contributions of fiberglass fabrics combined with ease of fabrication of composites has resulted in the acceptance of composites in many industries. Many aerospace and aircraft components, ranging from radomes, wing-to-fuselage fairings, and wing flaps on the exterior of aircraft to entire cabin and cargo-hold interior sidewalls are made of fiberglass-fabric-reinforced thermosetting resin composites, where high

*Reprinted by permission of Uniglass Industries.

strength, low weight, and part reliability are necessary design criteria.

Fiberglass fabrics also find extensive applications in recreational consumer products, among others, fishing rods, golf club shafts, vaulting poles, skis, bows and arrows, protective helmets, hockey sticks, and antennas. These articles all make use of fiberglass fabric's contribution to strong lightweight flexible products.

The use of fiberglass fabrics in the manufacture of high-strength, corrosion-resistant pipe has the advantage that large sections of pipe can be quickly made compared to other pipe-manufacturing processes. In addition, woven fiberglass products continue to find widespread application in boats, mariner docks, and tooling for many transportation industries.

Coated Fabrics

Fiberglass fabrics are coated with a variety of thermoplastic resins, rubbers, or elastomers for use in flexible air ducts, air-supported structures, conveyer belts, welding curtains, ceiling board facings, gaskets, pipe jacketing, pressure-sensitive tape, window shades, movie screens, and the like.

Scrims

These lightweight open-mesh fiberglass fabrics are used extensively in the construction industry as roofing membranes and wallboard-sealing tapes and masonry-crack repair.

Filtration

The thermal properties of fiberglass fabrics make them suited as high-temperature filtration media, especially with the pressure upon industry to eliminate air pollution in the environment. Baghouses equipped with fiberglass-fabric filtration bags are efficiently removing product and/or particulate effluent from exhaust gases in cement, carbon black, steel, nonferrous metal, and coal-fired boiler and power industries.

KEVLAR 49*

Kevlar 49 is the DuPont registered trademark for its new high-strength, high-modulus organic fiber. Combining high strength and

---

**Kevlar 49 Bulletin* (Wilmington: DuPont, 1977).

modulus with light weight and toughness, it is available in the form of yarns, rovings, nonwoven mats, and chopped fibers.

Yarn Properties

Density. Specific gravity of 1.44 is 40% lower than that of fiberglass.

Tensile strength. 525,000 psi. Equal to S glass and higher than graphite.

Modulus. 19×10^6; twice that of E glass.

Chemical resistance. highly resistant to organic solvents, fuels, and lubricants.

Textile processability. Excellent. Can be readily woven on conventional fabric looms. Yarns retain 90% of tensile strength after weaving. Can be easily handled on filament-winding and pultrusion equipment.

Flammability. Inherently flame resistant. Does not melt.

Temperature resistance. No degradation of yarn properties in short-term exposure up to 500°F.

Epoxy Composites

Density. 25–30% lower than that of glass composites.

Specific tensile. Superior to other composites reinforced with commercially available fibers.

Specific modulus. About twice that of S-glass composites.

Processability. Can be readily handled using conventional reinforced-plastics fabrication techniques.

Environmental stability. Useful long-term temperature range: −320°F to 320°F.

Flammability. Meets all FAA flammability and smoke requirements in flame-resistant, low-smoke resins.

Impact strength. Far exceeds impact strengths of graphite and boron.

Specific compressive strength. About half that of glass composites.

9 Compression and Transfer Molding

COMPRESSION MOLDING

Compression molding is the process in which thermoset molding compounds are formed and cured in matched metal molds under heat and pressure. The pressure, and often the heat, are supplied by compression presses to which the molds are fastened during the molding operation. Some thermoplastics are also compression molded but, by and large, compression molding is a thermoset process.

Molds

Most compression molds consist of two blocks of hardened steel that have been machined in such a manner so that when assembled the matching surfaces form the outline of the desired end product. The top section is called the *force* (or plug) and the bottom section is called the *cavity*. After the molding compound has been placed in the cavity, the mold is closed, and the part is formed in the hollow space created by the mating surfaces. Mold surfaces are either highly polished or chrome-plated in order to obtain smooth surfaces in the molded part.

Thermosets are adhesive by nature so that improper preparation of the mold surfaces can present a problem. Release agents such as silicone and high-temperature wax are applied to the contact surfaces to prevent sticking. Chrome surfaces also contribute to good release action.

Compression Presses

Most presses used for compression molding are actuated by a hydraulic ram moving either upward or downward, up-acting presses being more common. (See Figures 9-1 and 9-2.) Presses consist of two flat surfaces called *platens,* one of which remains stationary during the molding

Figure 9-1. 300 ton compression press. *(Courtesy of Dake Corporation)*

operation while the other moves up and down along four-corner posts. The force is fastened to the top platen, and the cavity to the bottom platen. The stationary platen is manually adustable up or down. Adjustment is made to provide the correct *daylight opening,* which is the distance between the platens when the press is fully opened. The opening should be just enough to permit removal of the molded part. Excessive daylight distance extends the ram travel, which would be undesirable because once the molding material has been placed in the cavity, the mold should be closed as soon as possible to avoid pregelation of the material in contact with the cavity surface.

Fastening the Mold

Press platens usually have tapped holes that are used to fasten down the mold. The assembled mold is placed in the center of the bottom platen; then the press is closed so that the top platen engages the top of the mold. Next, the mold halves are secured by bolts and clamps. Some platens have T-slots milled into them. For this type, molds are fastened with matching T-bolts.

Closing and Opening the Mold

After the molding material, called *charge,* has been placed in the cavity, the mold is closed fairly rapidly until the mold halves begin to telescope, at which time the speed is slowed considerably to allow the material to be properly heated before application of full pressure. With some materials, a hurried final closing usually results in a scrapped part and, in some cases, a damaged mold. *Guide pins,* usually mounted in the force, engage the guide-pin recesses, or *bushings,* in the cavity to assure perfect alignment when the mold is closed. *Knockout pins* are used to eject the cured part from the mold. They can be installed in either the force or the cavity and are automatically actuated when the mold is opened.

Figure 9-2. 2,000 ton compression press. *(Courtesy of Erie Press Systems)*

Heating the Mold

Most platens are heated by electricity or steam. High-pressure steam assures quick, uniform heating, but electrical cartridges mounted in the platens are more popular because they provide higher temperatures and cleaner operation. Small molds can be heated by conduction of heat from the platens, but deep molds must be heated internally to achieve adequate, uniform heat. To minimize the loss of heat, asbestos fiberboard or other insulation is located between the platen and platen mounting *(bolster plate)*.

TYPES OF COMPRESSION MOLDING

There are two distinct types of compression molding. One type involves the use of advanced B-stage molding compounds in the form of firm (or hard) dry powders, granules, chopped fibers, and diced or macerated fabrics. To be properly molded, this type requires high pressures. The other type consists of soft and/or pliable materials that do not require high pressures primarily because they are in an uncured or slightly advanced condition. This group includes premix molding, mat and preform molding, and sheet molding compound (SMC). Most thermoset molding in metal molds has been variously described as: (1) matched metal molding; (2) matched die molding; and (3) compression molding. All these terms however, describe a process that involves compression molding in matched dies (metal molds). For this reason, the categories can be simplified to "high-pressure compression molding" and "low-pressure compression molding." Most well-advanced molding materials are cured in excess of 3,000 psi. Soft materials are cured between 50 and 1,500 psi, with most cured under 1,000 PSI. Some low-pressure molding compounds are firm or hard, but most can be described as soft or pliable. As a general rule, high pressures are those in excess of 1,000 psi.

HIGH-PRESSURE COMPRESSION MOLDING

Molding Materials

Powders, granules, and chopped fibers are preferred for deep draw (tall) parts because of their superior flow properties. Most diced fabrics are used primarily for smaller sections because of their limited flowability. They do, however, provide the strongest molded parts, especially those made with bidirectional fiberglass. Somewhat lower pressures must be used when molding high silica compounds because of their inherently low crush resistance.

General-purpose phenolics such as those that are woodflour-filled are very popular with compression molders. Their low cost and outstanding moldability make them very attractive for both compression and transfer molding, and as powders and granules they are easily preformed on automatic preformers. The one drawback in molding with phenolics is their limited colorability, and when coloring is important, the molder usually selects melamine, urea, or polyester, which can be molded in a wide range of colors and shades.

Silicone molding materials are used for their resistance to high temperatures. Most are resistant to 600°F and some even higher. Phenylsilanes and polyimides are also used in high temperature applications.

Most thermoset materials are molded at 250–350°F. Deep, thick parts are usually molded at the low end of the temperature range to allow the material to heat and gel uniformly. Thin parts can be molded at higher temperatures and, obviously, much more quickly.

When thermoplastics are compression molded, they must be cooled below their heat distortion point before they can be removed from the mold. Thermosets are ejected immediately following the cure cycle.

Bulk Factor. This is the ratio between the volume of the molding material and the volume of the finished part.

$$\text{Bulk factor} = \frac{\text{Volume of molding material}}{\text{Volume of finished part}}$$

Low bulk factor materials such as most granules and powders usually fit into the mold cavity with no problem. They are often compressed into smaller volumes to gain other advantages. Fluffy, shredded materials have a high bulk factor and must usually be preformed to fit into the mold cavity. They do not lend themselves well to automatic preforming but can be adequately preformed by hand with an arbor press.

Preforms and Preheating. Preforming involves the reduction in volume of a molding material so that it will be easier to handle, easier to work with, and easier to preheat. The molding compound is compressed into small, convenient configurations. Preforms are especially useful when feeding multicavity molds. A fixture is positioned over the loading well, and the preforms are released into each cavity simultaneously. The use of preforms allows the mold designer to minimize the size of the mold.

Preheating of B-stage molding compounds should be included in the molding operation whenever practical. The combination of dense preforms and electronic (high-frequency) preheating results in reduced pressure requirement, shorter cure cycles, and higher strength in the cured part. Dimensional stability is also improved. With this method, the preforms are uniformly heated close to the required molding temperature in a matter of seconds. In the absence of an elctronic heater, a standard air-circulating oven can be used. Because the material heats much more slowly with this method, care must be exercised not to advance the material. Also, when preforms are not used, the material should be rotated occasionally to prevent uneven heating. Covering the material loosely with perforated aluminum foil can be helpful.

Degassing. When molding phenolics, the mold must be opened just a crack for a few seconds after it is first closed to allow for escape of trapped gases. This is known as *breathing the mold.* Phenolics give off water during cure. A failure to breathe the mold usually results in gas pockets and a scrapped part. Epoxies and polyesters do not give off by-products, but like many compression molding materials they do contain some volatile matter. For these materials the clearance between the force and the cavity is usually adequate in providing an escape for trapped air and volatiles. When the clearance is not quite enough, several vents 0.002 in.–0.003 in. × ¼ in. (to ½ in. on the land areas) usually remedy the situation. All vents must lead to the atmosphere.

Molding Sequence

1. Compute the amount of molding material needed. This is called *weight of charge.*
2. Preheat.
3. Apply mold release to molding surfaces.
4. Place the material in the cavity or cavities.

5. Close the mold.
6. Degas, if needed.
7. Open the mold at the conclusion of the cure cycle and eject the part.
8. Place the part on a shrink fixture, if available.
9. Blow flash off the mold with an air gun and apply more release agent when needed.

Molding Pressure

Pressure setting for B-stage compression molding compounds can be obtained by using the following formula:

$$\text{Pressure (tons)} = \frac{2{,}500 \text{ psi} \times \text{projected area (sq in.)}}{2{,}000 \text{ lb}}$$

This formula is for shallow parts. For deeper parts, add 500 psi for each inch of depth after the first inch.

Sample Problem: Calculate the pressure requirement to mold a part 10 in. × 5 in. × 6 in. deep.

Solution:

$$\frac{2{,}500 \text{ psi} + (500 \times 5) \times 50}{2{,}000} = 125 \text{ tons}$$

added pressure (psi)	500 × 5 = 2,500 psi
total pressure required (psi)	2,500 + 2,500 = 5,000 psi
projected area (sq in.)	10 × 5 = 50 sq in.
guage setting	5,000 × 50 ÷ 2,000 = 125 tons

When computing the projected area, be sure to include the land area. As noted, this area is pressurized at the same rate as the molding compound.

Estimating the Weight of Material Needed

The weight of charge can be estimated when the specific gravity and the volume of the part are known. Specific gravity is supplied by the vendor and the volume (cu in.) can be computed by multiplying the area (sq in.) by the thickness (in.).

$$\text{Wt of charge (lb)} = \text{sp gr} \times 0.036 \times \text{vol (cu in.)}$$

For small parts weighing less than a pound (454 gr), the following formula can be used:

$$\text{Wt of charge (grams)} = \text{sp gr} \times \text{vol (cc)}$$

where 1 cu in. equals 16.4 cc

Computing the area of parts having highly irregular contours can be very time-consuming. One way to simplify the calculations is to draw an outline of the part on a piece of paper and move and relocate protrusions until you have formed a symmetrical shape such as square, rectangle, or circle.

Types of Molds

There are three types of compression molds: positive, semipositive, and flash. A *positive mold* exerts full pressure of the force directly on the

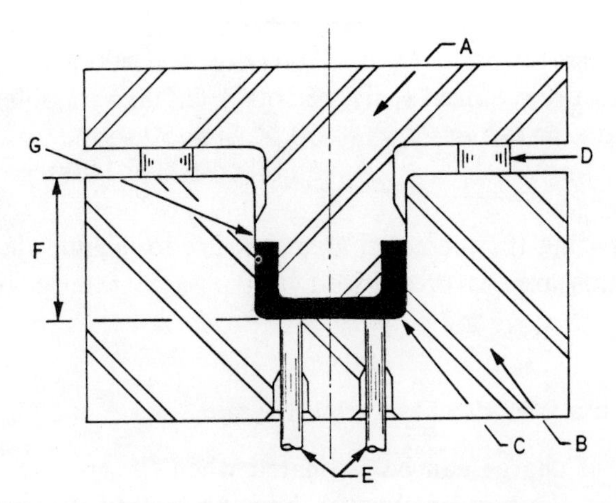

Figure 9-3. This illustrates a positive type compression mold. (A) Plunger or force. (B) Cavity. (C) Plastic part. (D) Guide pins. (E) Knockout pins. (F) Loading space. (G) Vertical flash ring.

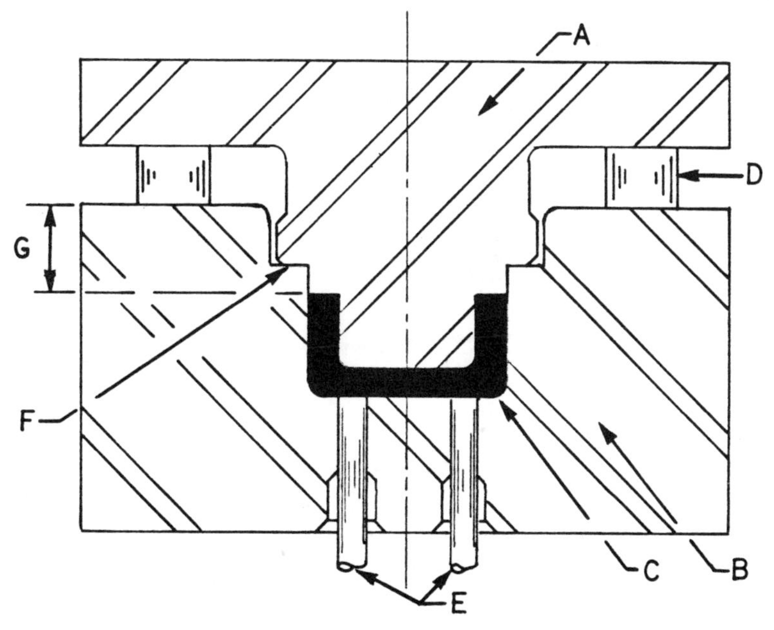

Figure 9-4. This illustrates a semipositive compression mold with vertical flash. (A) Force or plunger. (B) Cavity. (C) Plastic part. (D) Guide pins. (E) Knockout pins. (F) Land. (G) Loading well.

molding material. This type provides the densest moldings, but the weight of charge must be very accurate because there is little or no provision for escape of excess material (Figure 9-3). *Semipositive molds* (Figure 9-4) are so named because some of the pressure of the force is exerted on land areas of the cavity section during the cure cycle. The weight of charge must be reasonably accurate even though some material does escape when the mold is closed. Semipositive molds, which are used extensively in compression molding, provide good dimensional control and density. *Flash molds* are designed to operate with an overcharge of material. When the mold closes, all excess is squeezed out. They are less costly to make than other types but density and dimensional uniformity are not as good. They are used primarily for shallow parts (Figure 9-5).

Advantages of Compression Molding

1. There is a minimum amount of wasted material and there are no culls or runners to throw away.

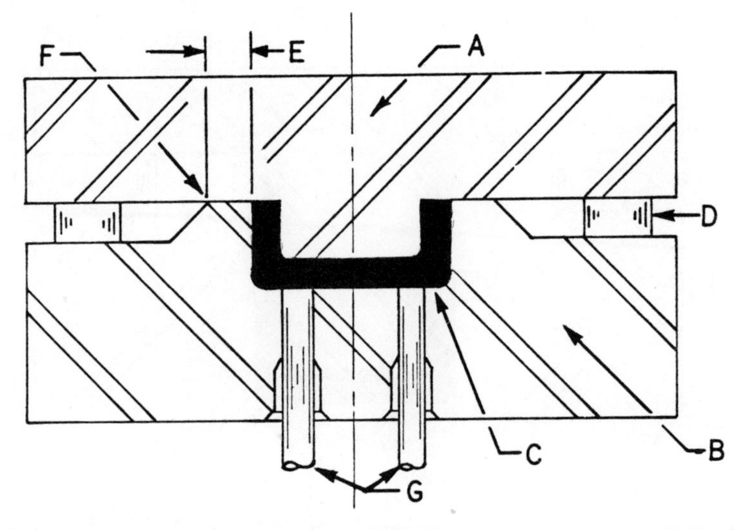

Figure 9-5. A flash type compression mold. (A) Plunger or force. (B) Cavity. (C) Plastic part. (D) Guide pins. (E) Land area. (F) Parting line. (G) Knockout pins.

2. There is no erosion of gates or runners.
3. As the pressure is applied more evenly, the internal stresses and resultant warpage are less.
4. Automatic equipment is available for cold powder, high frequency preheated powder, and extruded, preheated slugs.
5. Greatest dimensional accuracy is possible with the use of landed full-positive molds.
6. Shrinkage is less and reproducible, leading to more accuracy (i.e., more predictable shrinkage).
7. In general, thicker sections can be molded easier with compression than with transfer.
8. More cavities per mold are possible as lower pressure is required than transfer.

Limitations of Compression Molding

1. The process is not advisable with delicate inserts or mold sections.
2. Cleaning inserts is very often a problem.
3. Uneven parting lines can be a mold-design problem.
4. Flash tends to be thicker.

5. Cleaning of flash on high-impact materials can be time-consuming and a costly operation.
6. Molded-in holes limited to 2½ times the diameter.
7. Shot weight control is crucial on full-positive molds.

SOLVING MOLDING PROBLEMS (HIGH-PRESSURE MOLDING)

Mold Sticking

1. Check for proper coverage of release agent.
2. Preheat to eliminate moisture.
3. Raise mold temperature.
4. Increase cure.
5. Check mold material with small lab mold.

Dull Surface

1. Raise mold temperature.
2. Polish mold.

Blistering

1. Preheat.
2. Increase preform temperature.
3. Breathe the mold (twice, if necessary).
4. Lower mold temperature.
5. Increase cure time.
6. Check mold for proper venting.

Warping

1. Lower mold temperature.
2. Increase cure time.
3. Preheat.
4. Use shrink fixture.

Cracking

1. Check knockout pins.
2. Use more flexible material.

3. Check size of shrink fixture.
4. Lower cure temperature; extend cure time.

Unfilled Mold

1. Preheat to highest tolerable temperature.
2. Close mold faster.
3. Lower mold temperature.
4. Check for correct charge weight.

Weak Moldings

1. Check for correct weight of charge.
2. Increase cure time.
3. High-frequency preheat.
4. Postcure.

Orange Peel

1. Close slowly on low pressure.
2. High-frequency preheat.
3. Lower mold temperature.

Pitted Surface

1. Preheat.
2. Use stiffer material.

Burn Marks

1. Reduce mold temperature.
2. Preheat.
3. Close mold faster.

Poor Electrical Properties

1. Preheat.
2. Reduce temperature.
3. Increase cure time.

TRANSFER MOLDING

Transfer molding is similar to compression molding except that instead of the molding material being pressured in the cavity, it is pressured in a separate pot, then forced through an opening and into a closed mold. Transfer molds normally have multicavities. The molding material, under heat and pressure, reaches the cavities via runners and gates—the same as in injection molding. Pot transfer molds consist of three main sections: the bottom section, which houses the mold cavities; the center section, or material pot, which is clamped to the bottom section; and the top section, or plug, which pressurizes the molding material (Figure 9-6).

Transfer-molding pressures should be figured on the basis of the total projected mold area, including the part, gate, runner, and pot area. An extra 10% clamping pressure should be added to prevent the mold from opening and flashing. The transfer mold surfaces should be relieved rather than have flat mating surfaces so that if flashing does occur the mold will reclose after the transfer cycle. Lands of 3/16 to 1/4 in. are sufficient, and landing pads should be incorporated in the mold to hold open 0.0005 in. when closed to prevent damage to the lands.

In general, the studies of gate and runner design have shown the following facts about the effects of their design on the transfer-molding process:

1. Full-round runners offer less resistance to flow than half-round, square, or rectangular runners of the same cross-sectional area. Sharp corners in runners and gates can cause hot spots where materials set prematurely or burn before flowing into the cavities.
2. Three closely grouped gates do not fill the cavity faster than a single gate.
3. Multiple gates can produce knit lines where the material flows together. Usually one larger gate is better than several small ones.
4. For a given filling time, an 8-in.-long runner requires twice as much pressure as one of the same diameter and 2-in. long. The shortest possible runners should be used. Runners of equal length allow for balanced flow into the cavities and better flash control.
5. As thermosets are forced to flow rapidly through small gates, the pressure drop is converted to heat in the material. The heat speeds the cure and shortens the cycle. Thus, where high transfer pressures are tolerable, small gates are usually advantageous.

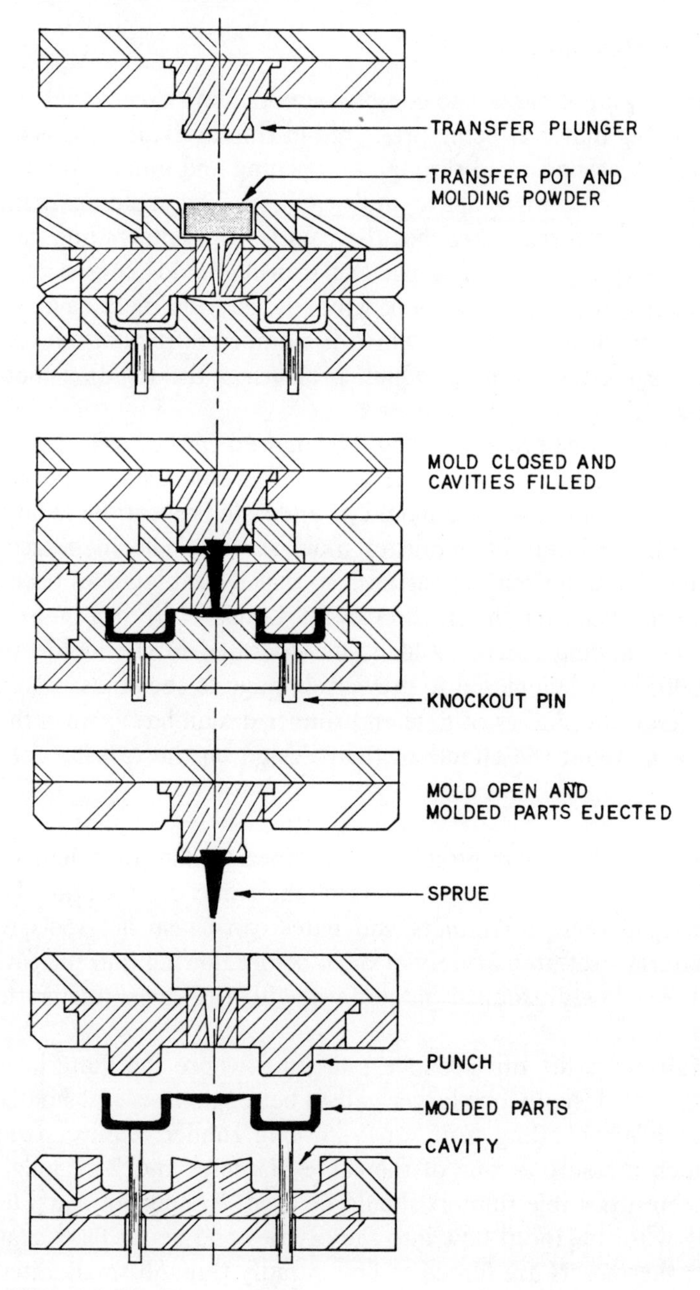

Figure 9-6. This illustrates the pot or sprue type transfer molding of thermosets.

Shrinkage in transfer molding is from 25 to 50% greater than in compression molding. For example, there is a difference of 0.004 in./in. between 2,500 and 10,000 psi.

Plunger Molding

A study of the diagram of a plunger mold (Figure 9-7) shows clearly the difference between transfer (pot) molding and plunger molding. Basically, they are the same except instead of three main sections, plunger molds have only two. Also, instead of feeding the material into a pot, preforms are simply dropped onto the mold through an opening in the top section, then pressurized by an indepently controlled, or auxiliary, ram. As in transfer molding, the material is forced through runners and gates and into the mold cavities. As noted in Figure 9-7, cull, runners, and molded parts are ejected intact as a single unit. In a pot transfer mold the plug is undercut so that cull and sprue are both retained when the mold is opened. (The sprue is the material formed in the tapered sprue bushing of a transfer mold. Cull is the cured excess material left in the bottom of the feed chamber of either type mold.) The thickness of the cull has little bearing on the molded part, since the material enters a closed mold. Economy, however, dictates that culls be held to a minimum. The cull does have to be slightly thicker than the runners to insure continuous pressure.

Fully automated compression presses for plunger molding are now providing cycle speeds approximating those attained with injection molding.

Advantages of Transfer Molding

1. It provides shorter cycles due to more heat of friction, and better heat transfer. Loading of preforms into the pot is less time consuming than loading preforms into each cavity.
2. In general, tool maintenance costs are lower. Wear of the cavities is less, but gates and runners erode.
3. Longer pins can be used, and they may be of smaller diameter as they can be telescoped or supported on both ends.
4. Delicate inserts and mold sections can be used as the mold is already closed before molding.

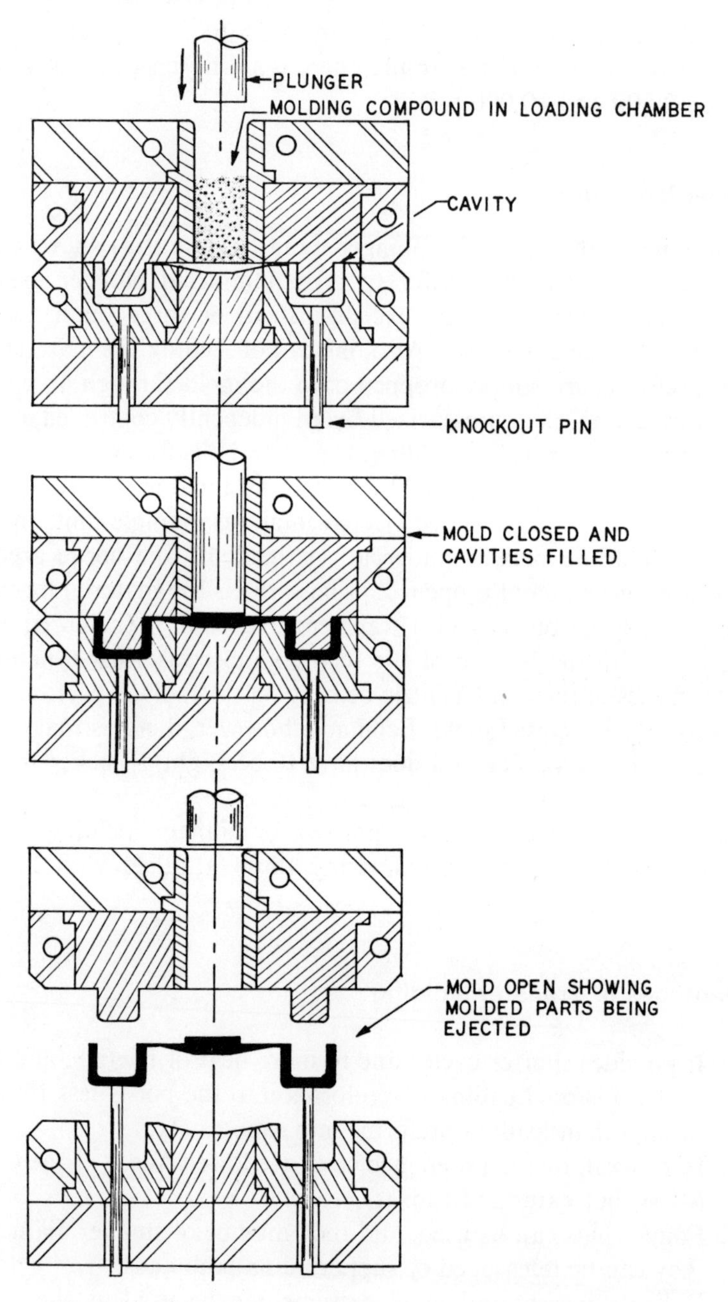

Figure 9-7. This explains the plunger transfer molding of thermosets.

5. It is possible to achieve essentially flash-free parts and flash-free insert faces because the insert can be clamped between the two halves of the mold.
6. Tighter tolerances perpendicular to the parting line are possible. In a properly designed and operated mold there is extremely thin flash.
7. Higher tensile and flexural strengths are obtained than in compression, whereas impact is lowered. Values vary with the individual material, but if we take compression-molded specimens as 100%, transfer values will run 80% for impact, 115% for tensile, and 120% for flexure. The same values are obtained regardless of method of transfer—conventional, two-stage screw, or in-line screw.

Limitations of Transfer Molding

1. Warpage occurs due to uneven filling especially with an unbalanced mold layout.
2. Knit lines appear in back of pins and inserts.
3. Wasted material results due to the cull and runner system.
4. Higher pressures are required, so fewer cavities can be put into a press of the same tonnage. Pressures of 8,000–15,000 psi are needed for transfer molding B-stage molding compounds.
5. Gate removal is necessary unless a pin-point or tunnel gate is used.

A picture of a transfer press can be seen in Figure 9-8.

LOW-PRESSURE COMPRESSION MOLDING

Premix Molding

Molders usually prepare their own premix molding compounds. The materials are mechanically mixed, then refrigerated in polyethylene bags until ready for use. Batch sizes are regulated so that all the material will be used while still in a moldable condition.

Most premix formulations have a soft puttylike consistency and are composed of polyester resins, catalysts such as benzoyl peroxide (BPO), fillers such as clay and calcium carbonate, and short fibers such as

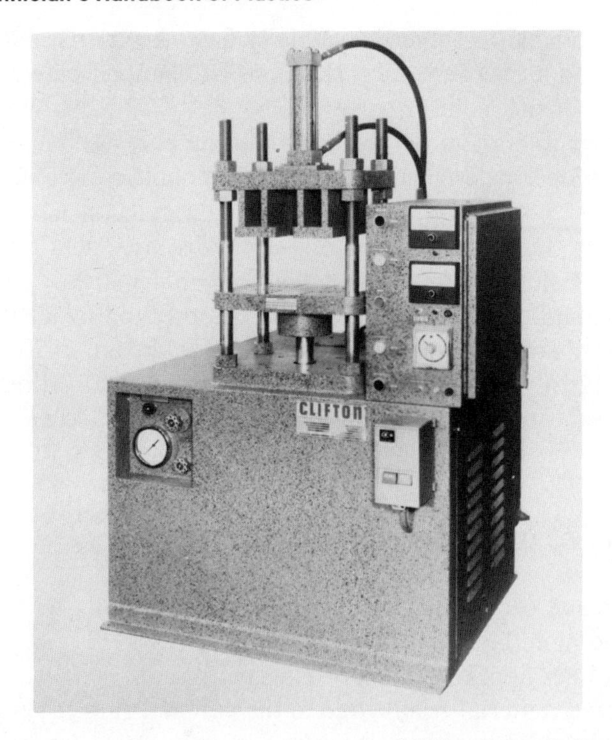

Figure 9-8. Hydraulic-transfer molding press. *(Courtesy of Clifton Hydraulic Press Company)*

fiberglass, asbestos, nylon, and sisal. A high filler content helps to provide good flow characteristics by holding all the ingredients uniformly intact. Asbestos also contributes to good flow stability but cannot be used in all formulations because of its poor coloring properties. (The danger of exposure to asbestos fiber must also be taken into consideration.) Special formulations featuring less filler and more and longer fibers have more of the look and feel of a fibrous mass. Release agents such as zinc stearate are used in all formulations to reduce internal friction.

Premixes, including some highly specialized compounds, are also available for purchase. They are marketed as pliable bulk, flake, and continuous rope.

A typical cure cycle for a standard premix molding compound would be 300°F and 800 psi pressure for 45 sec. A shrink fixture is normally used for premix-molded parts. The part is pulled hot and placed on the shrink fixture until cooled to ambient or near-ambient temperature.

Small shrink fixtures can be made easily by casting a mix of liquid epoxy, Versamid 125 or 140, and fine aluminum powder into a model part. The mix should be 70 parts epoxy and 30 parts Versamid 140, or 60 parts epoxy and 40 parts Versamid 125. Aluminum powder is added until the mix is thick but still readily pourable. Cure can be carried out at room temperature for 24 hours or at 130–150°F for 2–4 hours.

For large parts, the mix can be thickened with powdered silica so that the fixture can be made hollow. Instead of pouring the mix, it is troweled onto the model walls. This type of casting should be cured at room temperature to prevent thinning of the resin. Even at room temperature the casting should be checked periodically for sagging.

Premixes can also be molded by transfer and injection methods. With the advent of low-profile, low-shrinkage systems has come a rise in interest in premix molding. The smooth surfaces obtained with these systems plus the low overall molding cost and short cure cycles account for the steady rise in premix molding. The high filler content of premixes contributes substantially to the comparatively low cost of premix molding. Low profile (smooth surface) and low shrinkage are brought about by modifying the resin with a thermoplastic powder.

Chemical thickening agents are added to some premix compounds. They produce a tough, superficially gelled condition that results in more uniform physical properties.

Preform Molding

This method utilizes polyester-impregnated fiberglass preforms as the mold material. Not to be confused with the small, dense preforms used in high-pressure molding, these preforms are composed either of randomly oriented continuous glass fibers with or without a binder or of chopped strands held together with a binder. For both types the fiberglass is built up on a form that closely approximates the shape of the part to be molded. The handleable preform is then placed in the mold, either on the force or in the cavity. A measured amount of catalyzed resin is then placed over it, and the preform is ready for pressing into the cured end product.

Chopped strand forms are prepared by blowing short fibers over a shaped metal screen. Air suction is used to disperse the fibers uniformly over the screen. When the form is complete, it is sprayed with a liquid binder. After the binder has been heat-cured, the preform becomes handleable and can be transferred to the mold. Sometimes the cured part can be ejected easier from one half of the mold than it can from the other half. It can be induced to remain in the section from which it is easier to remove by creating a temperature differential between the two mold halves. A 5–10° drop in either half usually keeps the part in that half.

Fillers such as clay and calcium carbonate, which are also used in premix compounds, are usually added to the catalyzed resin. As with premixes, they improve flow properties and surfaces and lower costs.

Pressures used for preform molding are generally under 500 psi.

Preform-molded parts are stronger than premix-molded parts because of the greater concentration of fiberglass. Also, addition of the resin at the press eliminates the possibility of pregelation. With all other thermoset molding compounds, because the materials are mixed well in advance of the molding operations, care must be used to prevent premature advancement of the resin. For some thermoset processes, inhibitors are added to polyester resins to prevent pregelation.

Smc Molding

This is one of the newest and most promising material forms being used in reinforced polyester molding. SMC (sheet molding compound) is a continuous sheet of molding material that has been made firm and handleable by incorporating chemical thickening agents into the mix. When ready for use, pieces are simply cut off the roll and placed in the mold cavity.

The molding sheets consist of catalyzed resin, fillers (including thickeners), and chopped glass fibers. Low-shrink resin systems are also popular with SMC molders. The low-profile, low-shrink characteristics are brought about by the addition of a thermoplastic powder to the polyester resin. As previously noted, premix molders also use a similar low-shrink system. "Low-profile" refers to the surface smoothness obtained with low-shrink systems.

Because of the thickened consistency of SMC, molding pressures are higher than for premix, mat, and preform molding. It is anticipated that improved compounding will soon allow SMC materials to be molded at pressures considerably below those now being used.

SMC molding requires an ejector pin-system in the mold.

MOLDING PROBLEMS (POLYESTERS)

I. Exposed Fibers
   A. Too much mold lubricant
   B. Mold too hot
   C. Glass content too high
   D. Poor glass distribution
   E. Coarse preform (use surface mat or finer glass fibers)
   F. Character of resins (those that distort at too low temperature)
   G. Improper molding temperature
      1. Raise or lower temperature
      2. Use temperature differential on halves and it will reduce fiber and increase the glass on the hotter side
II. Color Variation
   A. Poor pigment mixing
   B. Poor glass distribution
   C. Nonuniform cure of preform binder
III. Delamination
   A. Poor impregnation
   B. Glass content too high
IV. Styrene Odor
   A. Undercure
   B. Additive used that is inhibiting the cure
V. Warping
   A. Unbalanced construction of resin and glass
      1. Use less styrene
      2. Use filler
      3. Use lower molding temperature
   B. Uneven cure
      1. Adjust mold surfaces to same temperature
   C. Design
      1. Enlarge (open up) radii
      2. Stiffen (build up) sides
      3. Use resin with higher heat distortion temperature
VI. Washing of Preform
   A. Poor preform
      1. Loose fibers with insufficient binder
      2. Using binder that is soluble in the resin and has dissolved before the resin flow has ceased

- B. High viscosity resin
  1. Too much resistance to flow, add styrene, reduce filler use a filler with a lower oil number
- C. Early gel before press closes
  1. Reduce catalyst, add cure inhibitor

VII. Blisters
- A. Mold too hot
- B. Moisture in resin or filler or trapped on glass
- C. Cure too rapid
- D. Undercure
- E. Entrapped air
- F. Expanding vapor (moisture or solvents)

VIII. Pinholes, Pits, Voids
- A. Dirty mold
- B. Poor (porous) mold surface
- C. Entrapped air in resin
  1. Mixing too fast
  2. Insufficient pressure
  3. Lower temperature

IX. Crazing
- A. High exothermic reaction
  1. Reduce catalyst
  2. Reduce mold temperature
  3. Add filler, add flexible resin
  4. Use less styrene in resin
- B. Resin-rich areas
  1. Poor glass distribution
  2. Poor part design
- C. Undercure

X. Sink Marks
- A. Cure too rapid
- B. Mold too hot
- C. Insufficient pressure
- D. Poor part design
  1. Change in thickness of the part too rapid

XI. Ripples and Wrinkles
- A. Preform too large
- B. Cure too rapid
- C. Use of old resin

D. Improper placement of reinforcements

E. Washing (movement) of glass

XII. Rips in Glass Fibers

A. Preform too large

REFERENCES

Technical Literature No. 350. Durez/Hooker, North Tonawanda, N.Y., 1979.

Milby, R. V. *Plastics Technology*, McGraw-Hill, New York, 1973.

Mohr, J. G., Oleesky, S. S., and Shook, G. D. *SPI Handbook of Technology and Engineering of Reinforced Plastics/Composites*. Van Nostrand Reinhold Co., New York, 1973.

Beck, R. D. *Plastics Product Design*. Van Nostrand Reinhold Co., New York, 1970.

10 Hand Layup

This is the process in which layers of pre-preg or freshly wet fabric or mat are placed on a single mold and cured to a solid state. The shape of the mold dictates the shape of the part. Hand layup is used when the size of the order is too small to warrant expenditure for matched metal molds or when the size of the part is too large for press work. Also, complicated shapes are easier to reproduce by hand layup.

Wet layup is one of the most common methods used in layup fabrication. Most wet layups are made with polyester resin and fiberglass, and they can be cured at room temperature or at an elevated temperature depending on the curing system used. For room-temperature cures a promoter such as cobalt naphthenate (cobalt) is first stirred into the resin then a catalyst such as methyl ethyl ketone peroxide (MEKP) is added. The heat of reaction (exotherm) caused by this mixture is sufficient to cure the resin, though some additional heat is often used to hasten the cure. Cobalt must never be allowed to come in direct contact with MEKP. When the two are mixed separately into a resin, there is no problem, but when they are mixed directly with each other, they are explosive.

Some wet layups are cured under heat and vacuum pressure. For this type layup a longer pot-life system is used to permit installation of a vacuum bag. Benzoyl peroxide (BPO) used in the proportion of 2% of the weight of resin and without a promoter is one of the most popular catalysts used for pressurized wet layups. Curing is carried out at 225–250°F.

Gel Coat

Most room-temperature-cured polyester layups include a gel coat that is applied to the bare mold surface before addition of the fiberglass. Gel coat consists of catalyzed resin, fillers, and usually a pigment. It provides a hard, smooth, glossy, colorful surface. After the gel coat reaches a near-cured or leathery state, it is coated with fresh catalyzed resin and covered with a specified number of fiberglass plies to complete the layup. If the fiberglass is applied before the gel coat reaches an advanced condition, the inclusion of fibers on the surface of the part becomes a distinct possibility. The plies are wetted out as needed. After addition of each ply, the surface is rolled or rubbed by a squeegee to minimize voids and assure intimate contact between plies.

Pre-preg layups are cured under heat and pressure with the pressure being supplied by means of a vacuum bag over the assembly. When the air is removed from the bag by means of the suction of a vacuum pump, the pressure of the atmosphere (approximately 14.7 psi) is exerted directly upon the bag and laminate. If more pressure is required, the assembly is cured in an autoclave at 50–200 psi. Because B-stage fabrics are partially cured, they require some pressure to laminate them properly, hence the use of vacuum and autoclave pressure. The resultant product is more void free, denser, and stronger than wet layups cured at room temperature with only contact pressure.

Types of Molds

Molds can be either a plug (male) or a cavity type (female). A cavity mold would be used to layup a boat because the outer surface of the boat must be smooth, glossy and colorful (function of the gel coat). A male mold would be used to layup a swimming pool because the inside surfaces must be smooth and glossy.

Plaster Molds. Most layup molds for small orders are made of plaster and are classified as temporary tooling. Plaster molds do not hold up well when used for high-temperature laminating. A master mold can be made, and a mold for each layup can be cast from it. If the layup mold is to be a male, the master mold would have to be a cavity type; if it is to be a female, the master mold must be a plug type. Molds for long runs are usually made of fiberglass-epoxy and are fabricated by one of the layup methods outlined in this chapter. Epoxies provide high interlaminar strength (bond between plies), good chemical resistance, low shrinkage, and hard surfaces. Metal and fiberglass-plastic molds are classified as Class A or permanent tooling.

Plastic Tooling. Used to make stretch, drop-hammer, and forming dies, trim and drill fixtures, etc., plastic tools provide time and cost savings over metals. Because they are laminated against smooth surfaces, they do not require any further finishing. In addition, there is no corrosion problem as with some metal tools. Redesigning is also simpler with plastic tooling. The surfaces can easily be recast or relaminated to comply with engineering changes.

A plaster mold can be made by taking a *splash* of the configuration to be reproduced, whether it be an actual part or a section of a part. Trim and drill fixtures for aircraft are often laid up on molds made in this manner. "Splash" is part of the plaster patternmaker's jargon that simply refers to the casting of plaster over a shape to be reproduced.

When there is no available configuration from which to make a mold, a replica can be built with plaster and *templates,* which are usually made from thin metal stock. The different profile shapes are laid out on a metal plate then carefully cut out and filed to net dimensions. They are assembled upright on a flat surface and bolted together along drill rods running horizontally through the templates. The mold is completed by sweeping on plaster until it is flush with the edges of the templates. The plaster is swept against the templates with a thin steel straightedge called a *screed.* (The sweeping operation is also known as *screeding.*) Layup molds with simple profiles can be made by passing a single shaped template across the top of wet plaster. The template is mounted on a "sled" and moves across the plaster in a straight line. Turntables can be used to make cylindrical molds. A mandrel with a crank at one end is attached to the top of an open wood frame. The plaster is fed onto the turning mandrel and brushes against a stationary template to form and shape the mold.

Undercuts. Care must be used when building a mold in order not to create undercuts that would make removal of the part impossible without damage to the part or the mold, or both. A male mold that is narrower across the lower areas than at the top areas is undercut; just the opposite proportions would create an undercut in a female mold. Molds that must include undercuts because of part configuration can be made with water-soluble plaster. When the layup is cured, the plaster mold is simply washed away with water. Hollow articles such as air ducts are also made with water-soluble plaster.

Mold Surfaces. Molds must have smooth, glossy surfaces in order to impart these characteristics to the finished part. A parting agent such as wax, silicone, or fluorocarbon is applied to the contact surfaces prior to the layup in order to provide good release action when the cured part is removed from the mold.

When making a wet layup on a female mold, better uniformity and control of resin content can be attained by impregnating the fabric before laying it on the mold. This will also help to prevent the formation of puddles caused by vertical drainage. Runoff on a male mold will flow off the layup so that wetting the glass on the mold does not present a problem. Mat is normally wetted on the mold to prevent unraveling and distorting.

Trimming the edges of a wet layup is much easier when the fabric is in a leathery or near-cure condition. A Stanley knife or similar tool will cut readily through the fabric at this time.

Low-pressure Lamination

Curing wet layups and B-stage fabrics by contact, vacuum, or autoclave pressure is called low-pressure laminating. High-pressure lamination involves the pressing of well advanced plies between the platens of a compression press at 1,000 psi or higher. Laminates that are compressionpressed with lesser advanced fabrics and lower pressures also fall under the category of low-pressure lamination. Figure 10-1 shows a multiplaten production laminating press.

VACUUM-BAG MOLDING OF PRE-PREGS

After the layup has been concluded, the assembly is covered with release cloth (silicone-coated glass cloth) or a perforated film. Release cloth is colored to distinguish it from other fiberglass fabrics. If a perforated film is used, it should be a nonstick type such as Teflon or Tedlar. In the absence of a nonstick film, any durable film will do, provided it is coated with silicone or another suitable release agent. A perforator can be made by driving finishing nails into a wood rod and filing the nail heads to a point. The roller is completed upon attachment of a handle.

After the release cloth or perforated film has been emplaced, it is covered completely by a porous bleeder material such as fiberglass, burlap, or cloth. The bleeder forms a passage for the exit of air and maintains continuity of the vacuum. Also, the bleeder absorbs all excess resin

Figure 10-1. Production laminating press. *(Courtesy of Clifton Hydraulic Press Company)*

that bleeds out of the layup. Special bleeder materials including some paper types are now commercially available.

The bag can be applied as either an overall envelope or it can be fastened down to the mold flange. Layups on porous molds without a base plate must be envelope-bagged to insure a complete seal.

Before a vacuum bag is installed, all sharp edges and protrusions must be padded to avoid puncturing of the bag.

Envelope Bag

To make an envelope bag, two pieces of film, usually square or rectangular and of equal size, are first sealed together on three sides. After the layup is wrapped or covered with a release ply and bleeder, it is placed inside the bag. The vacuum valve is then secured and the open end of the bag is sealed.

The layup is now ready for cure. The valve is positioned on a piece of bleeder material that is in direct contact with the layup bleeder.

Vacuum Valves. Vacuum valves are nothing more than hollow metal tubes with a rubber-lined flange on one end and a threaded area on the tube (stem) O.D., just above the flange. To install the valve, a small opening is made in the bag, and the valve is pushed through and into position. A rubber gasket and a metal plate are then placed over the assembly, and a nut is screwed on so that the bag membrane is secured between the two rubber surfaces. This prevents leakage during evacuation (pulling a vacuum to let the air out of the bag). Some molds for large layups have two or more permanent valves attached to the bottom of the mold base plate. This eliminates the necessity of installing valves in the bag and also provides for quick connect and disconnect. Vacuum valves must be kept open and resin-free at all times, which is one of the functions the bleeder serves.

Small layups are occasionally made without a valve. The end of the hose is covered with fiberglass or other porous fabric and inserted directly into the bag. The hose is sealed to the bag with zinc chromate or other sealant at the point of entry into the bag.

Sealing the Bag

Tape Sealers. Fastening the bag to the mold flange can be accomplished by one of several methods. The most common method involves the use of a sealing compound in a tape form. The sealant is applied around the mold flange, and the film is pressed securely into it. Excess bag slack is gathered in folds at several places, and the edges of the folds are filled with the sealant to make them air tight. Folds are known as "ears" in shop jargon. Before applying the sealant, the flange should be cleaned with acetone or MEK around the sealant contact area. The bag can also be fastened down by mechanical means, such as placing a rubber-lined metal frame over the flange and clamping in place. Thin, flat layups do not normally require the formation of "ears."

In the early days of layup, zinc chromate putty was used almost exclusively to seal vacuum bags. Since then, improved sealants less sticky and difficult to work with have gradually displaced zinc chromate in most layup shops. The newer sealants also remain stable at elevated temperatures. Vendors usually color the various sealant tapes to indicate temperature resistance. Black might indicate lower temperature use (to 200°F), while pink might mean to use a temperature up to

350°F. Zinc chromate putty is still utilized by some shops for small layups and for sealing bag leaks. It is actually an excellent sealant, but in addition to being difficult to work with it is also toxic and should be handled with care. Many laminators still refer to all bag sealants as "chromate," even though, in most cases, this is a misnomer.

Electronic Sealers. Electronic heat sealers are used extensively to seal bags. They are especially effective for large and very large layups, and the time-saving is appreciable. Seals made with electronic heat sealers insure a complete seal at first contact. Those made with rubbery or puttylike sealants against flange surfaces, although effective, depend more upon the thoroughness of pushing or squeezing the bag membrane into the sealant.

Nylon and PVA (polyvinyl alcohol) are the most common films used for vacuum bags, with nylon predominating because of its superior performance at elevated temperatures and autoclave pressures. PVA and PVC (polyvinyl chloride) are preferred for peripheral-bleeder, wet layups because of their superior toughness. Polyesters can be cured at 225°F, which is not excessive for PVA, but PVC should be used only for ambient- or low-temperature cures. Polyimide film, such as Du Pont's Kapton, is used for vacuum bags that are exposed to temperatures in the 500°F range.

All bagging films and other films stored on racks, including remnants, should be properly labeled. For example, Tedlar is a no-stick film but can also be prepared with bondable surfaces. It is difficult to distinguish the two kinds of films visually. Several other different types of clear films are also difficult to identify without a label.

When the bag has been secured, a partial vacuum is drawn so that the operator can still move the bag around to form the "ears" and gather slack in areas having sharp bends. If sufficient slack is not allowed in these areas, the bag will bridge rather than conform to the contours. After this operation, full vacuum is drawn, and the layup is ready for cure.

Thermocouple wires mounted in a trim end of the layup and connected to a recorder can be used to monitor actual part temperature. If the layup is made to net dimensions, a small separate layup can be made to monitor the temperature. It must be cured with the part layup in the same oven at the same time.

When a small bag leaks, a small gob of zinc chromate putty applied over the opening will usually seal it off. When there are several leaks, the chromate is applied all around the outside area of the openings, and a piece of bag film is secured over the entire area. When leaks are too numerous, a new bag may have to be installed, provided the resin has not started to gel. Vacuum gauges are mounted outside of the cure ovens so that pressure readings can be observed. In some shops, buzzers are used to warn of vacuum leaks. Leak-sensing devices are available, but most leaks can be heard by listening to the bag surface. A soft hissing sound is audible in the area of the leak.

The cured part should remain under vacuum during cooldown.

When working with tacky pre-pregs, it is sometimes difficult to remove the poly separator film. To make the task easier, apply dry ice to one corner and pull the separator away in a quick, jerking motion. Only a small area should be frozen to prevent moisture from condensing on the surface of the pre-preg. For the same reason, pre-pregs removed from the freezer should remain packaged until they reach room temperature. Cold-inducing sprays can also be used for freezing.

A small vacuum-bagged layup can be seen in Figure 10-2. Note the formation of the "ears" and location of the vacuum valve.

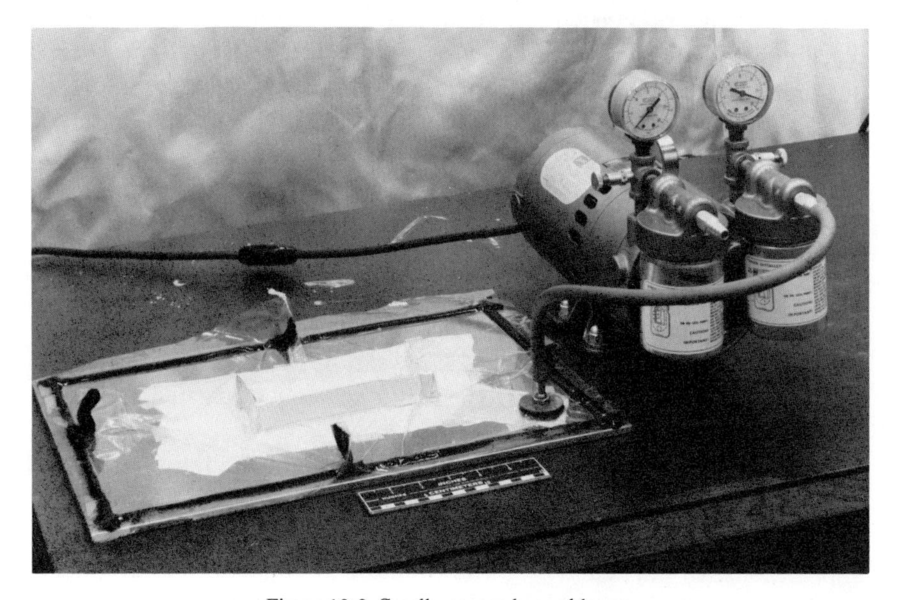

Figure 10-2. Small vacuum-bagged layup.

VACUUM BAG MOLDING OF POLYESTER WET LAYUPS

There are two methods for bagging wet polyester layups. One is identical to the method used for bagging pre-pregs, and the other method differs in the location of the bleeder material. Instead of the bleeder being applied over the entire layup, it is placed around the periphery and in contact with only the edges of the layup. This second method is used when both sides of the part are to be relatively smooth and air-free. The vacuum bag is installed directly over the layup. Because the bag is in intimate contact with the wet surface, it must be a nonstick type. If there is any doubt, a release agent can be applied to the contact surface of the bag.

After the vacuum has been drawn and the ears formed, a bleeding process is started. The outside of the bag is coated with a lubricating oil, and all air and excess resin are gradually forced into the bleeder material by dragging a squeegee across the bag surface in a center to edges pattern. Care must be exercised not to remove too much resin, which would result in dry areas. It is advisable to use an excess amount of resin for this kind of layup to be sure that there will be enough to squeegee away all the air.

A common practice with wet layups is to bleed, or complete the bleeding, after the layup has been in the oven for awhile—heat thins out the resin and simplifies the operation. Obviously, however, bleeding must take place before the start of gelation.

SANDWICH CONSTRUCTION

Sandwich construction consists of a lightweight core, such as honeycomb or foam sandwiched between two thin sheets of a rigid material such as aluminum, plastic, or wood. Though most honeycomb cores are weak materials and most facings *(skins)* can be easily bent, assembly of the three produces a thick, strong, stiff lightweight panel that would require a heavy load to break.

The high strength to weight ratios and the light weight of sandwich assemblies have made them an important part of aircraft and aerospace construction. Foam panels are being used in insulative and structural applications. In particular, rigid urethane foam appears to have a tremendous potential as a lightweight structural material of the future.

Sandwich Materials

Almost any sheet material can be used for facings, but the most widely employed materials are reinforced plastics and aluminum. The most frequently used materials for honeycomb cores are aluminum, fiberglass, paper, and cotton. Resin systems for sandwich construction include epoxies, phenolics, nylon-phenolics, silicones, and polyesters. High-temperature-resistant materials such as polyimides and polybenzamidazoles are used in specialized applications. Some of the newer exotic materials like high-strength graphite fabrics and the high-tensile-strength polyamides are also gaining increased use in aerospace honeycomb fabrication. The high-strength polyamides are being used for facings and core.

Among the considerations for selecting honeycomb core are material composition, core thickness, cell size, cell-wall thickness, and core density.

Aluminum Core. Aluminum core is available with either perforated or nonperforated cell walls. Resins such as phenolic give off gases as side products and cannot be used with nonperforated aluminum cores. Other resins, including epoxies, do not give off gases during cure, though pre-pregs prepared with these resins do contain small amounts of volatile matter in the form of moisture and residual solvent. While the vapors from these materials are normally carried away by the vacuum system, some fabricators prefer the perforations when working with aluminum core, regardless of the resin system being used.

Reinforced Plastic Facings. Most plastic skins for honeycomb sandwiches are made with pre-pregs. As with all pre-preg fabrication, there is an advantage over wet-layup methods in ease of handling, uniformity, and resin control.

Honeycomb sandwiches made with pre-pregs are usually assembled in one or two steps. In one-step assembly, the outer skin is laid up on the mandrel, and the core is positioned on it. The assembly is completed upon addition of the top plies that form the inner skin. The outer skin will be smoother than the inner skin because it is cured against the smooth, hard surface of the mold. The top, or inner, skin will show some indentation over the cells. Dimpling can be minimized by adding a thin, released stainless-steel caul plate before emplacement of the bleeder and vacuum bag. Pre-pregs for honeycomb construction are now designed to impart smoothness and high performance to the core facings with either one-step or two-step fabrication. The only adhesive used to bond to the core is the resin contained in the pre-preg. Built-in conditions of resin control assure strong, uniform bonds.

Pre-preg assemblies are cured at elevated temperatures under vacuum bag pressure. When more pressure is desired, the assembly is cured in an autoclave.

Wet-layup Honeycomb Sandwiches

Wet-layup honeycomb construction is usually processed in two or three steps. A three-stage layup can be processed as follows:

1. Lay up the outer skin on the mold and cure using standard wet-layup and vacuum-bagging techniques.
2. If a *peel-ply* ply was used on the laminate remove it at this point. When a peel ply is removed it leaves a rough, bondable surface free of parting agent and other contaminates.

   Apply an adhesive fillet around the edges of the core cells (one side) and place the core, adhesive side down, on the cured laminate. Place a thin layer of impregnated fabric over the top of the core and again bag and cure in the standard manner. The thin layer, called a *tie ply,* is used to close the cells to prevent resin from draining into them when the plies for the inner skin are added.
3. Apply the prescribed number of plies to the top of the cured assembly. Bag and cure. This forms the inner skin and completes construction of the sandwich.

Note: The edges of honeycomb assemblies that are cured under vacuum pressure must be protected from crushing or distorting. This can be accomplished by placing square or rectangular bar stock (of similar thickness) against all the edges before adding the vacuum bag and other accessory materials. The edge pieces should be wrapped in Teflon tape to prevent sticking to the core. Cores made of steel, titanium and other strong metals do not require edge protection.

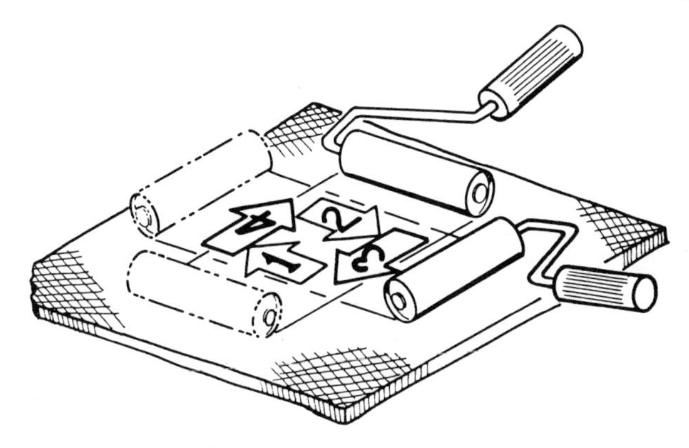

Figure 10-3. Rolling core for single box coat.

Applying the Corecoat

Most corecoat adhesive is applied by roller in four directions, as noted in Figure 10-3. This assures complete coverage of all cell edges. When the adhesive is not thick enough, it can be made thixotropic by the addition of silica powder. A good corecoat can be seen in Figure 10-4. Note that there is no sag after the coating has been applied.

Another method of applying corecoat to the cells is film transfer. The resin mix is applied in an even coating over a nonstick film such as Tedlar then transferred to the core. After gentle rubbing, the film is removed and weighed back to determine the exact amount of adhesive transfer. When the transferred amount is insufficient, the film can be placed back on the core and rubbed some more—or a second transfer can be made.

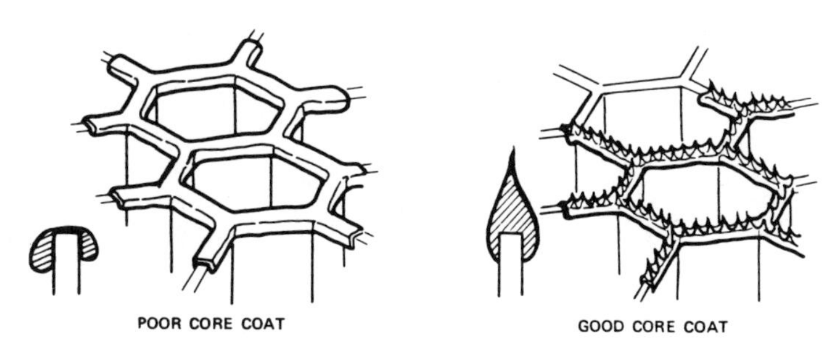

Figure 10-4. Core coat quality.

Honeycomb core, such as Hexcell's Hexabond, is now available with the adhesive already applied to the cells and ready for bonding.

Joining of Core Sections

Various methods are used to join together small core sections in order to form a larger production section. Fiberglass cores can be overlapped and pounded together with a mallet, and most cores can be spliced by simply bonding the edges together with an epoxy adhesive. The core can be held together by the use of small aluminum clips and adhesive. The adhesive can be a thin ribbon of pre-preg or a foam strip. When the adhesive has been laid on the clips are hooked on about every 2 in. In the case of the foam strip, the adhesive becomes activated upon addition of heat. Aluminum clips are especially helpful when applying the core to a vertical surface. This type of clip is not removed and remains as part of the assembly. Other clips, such as those shown in Figure 10-5, are Teflon-coated and are removed following the curing operation. Most adhesive foams are epoxies or modified epoxies with cure temperatures ranging up to 350° F. They are available in various widths that can be cut to size, and their thicknesses range from 0.05 to 0.1 in.

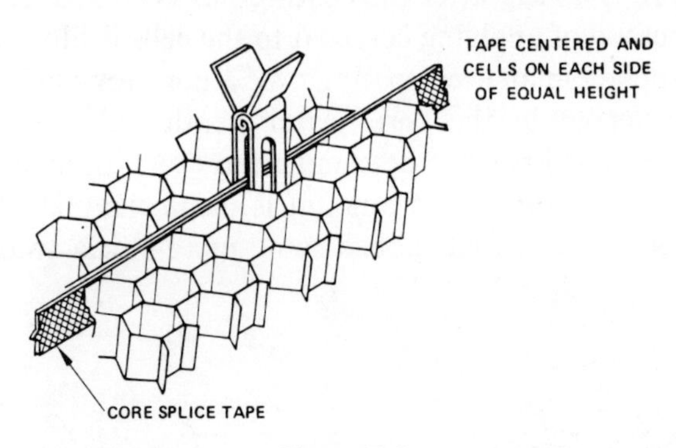

Figure 10-5.

11 Miscellaneous Thermoset Processes

MISCELLANEOUS THERMOSET PROCESSES

Potting, Encapsulation, and Casting

Potting and encapsulation have practially the same meaning insofar as both involve the pouring of a resin mix into a mold or walled area for the purpose of covering components with a protective barrier. While encapsulation is normally associated with closed molds and vacuum and pressure cycles, the term "potting" refers mainly to pourings made in open molds and cured with no pressure. The terms, however, are often used interchangeably around shops and labs.

Casting is another term used to describe the pouring and setting of a liquid mix, but, whereas potting and encapsulation involve the covering of components for protective purposes, casting refers primarily to the process of producing articles having a specific configuration. For potting and encapsulation, shape needs only to be convenient.

Casting Applications

Epoxy casting mixes are used in automotive and aircraft plants for molds and forming dies, and cast urethanes are used for bumper guards and other crash padding. The aerospace industry utilizes both epoxies and urethanes for protective encapsulation of electronic components. Silicone rubbers (RTVs) also have numerous aerospace applications.

Resilient polyester resins are finding ever-increasing use in the casting of furniture sections. By incorporating fillers such as wood and pecan flour, calcium carbonate, microspheres, fibers, and appropriate dyes, these castings are finished to look, feel, and work like real wood. The use of low-cost materials, low-cost tooling, room temperature curing plus a minimal floor space requirement, accounts for the growth and popularity of polyester furniture. Urethane is also used in making cast furniture sections. Polyester resins are widely used in the production of cultured-marble bathroom washbowls and tabletops. The castings contain approximately 80% natural marble in powder form and have the beauty of real marble without its porosity and brittleness.

Another large outlet for casting is in the production of molds. Silicone rubbers, in particular, because of their resistance to high temperatures (to 500°F), no-stick properties, and conformance to fine detail, are popular selections as materials for making molds. Of particular importance is the fact that they cure with little or no exothermic heat so that large sections can be cast without fear of degradation. Their no-stick property allows castings to be made in silicone molds without the use of a release agent; parts are easily removed simply by flexing the mold.

Epoxies, urethanes, plaster, and rubber latex are also used to make molds for casting.

FILAMENT WINDING

Filament winding is the process of winding a continuous filament, usually glass roving, around a cylindrical, conical or other shaped mandrel (mold) in the production of high-strength hollow structures. The aerospace industry uses filament winding to produce rocket motor cases and various types of pressure bottles. Commercial winders are engaged in the production of fishing poles, reinforced pipe, pressure vessels, and storage tanks.

Epoxy resin is used almost exclusively as the filament binder in aerospace windings, while commercial winders prefer the less expensive polyester resins. Though epoxies are generally superior in interlaminar strength, heat resistance, and cure-shrinkage, polyesters nevertheless do provide enough desirable properties to warrant their use in commercial manufacturing.

Winding Patterns

While under tension, the fibers are wound in a specific pattern. Orientation can be helical, circumferential, longitudinal, or a combination of patterns. For helical winding, the mandrel rotates continuously while the feed carriage moves back and forth at a controlled speed that determines the helical angle. For some windings, the feed arm revolves around a stationary mandrel.

Resin Systems

Resins for filament winding are normally low-viscosity systems under 1,000 centipoises. The thinner resins provide better wetting of the fibers and more thorough expulsion of trapped air. Another important consideration is the pot life of the resin, which should be of sufficient duration to insure completion of winding operations before the start of gelation.

Tape Wrapping

Tape wrapping is similar to filament winding except that, instead of using strands, the wrap is made with thin, flat, narrow pre-impregnated tape. The most common application of tape wrapping is in the production of rocket parts such as exit cones. The tape is spiral wrapped flat to the mandrel, except in *edge grain* applications where the tape is wrapped at an angle to the longitudinal axis of the mandrel. For tube wrapping, the tape, which is usually a pre-preg, is fed onto the mandrel under heat and tension. The heat and pressure serve to debulk the wrap and maintain good compaction. Cure pressure is usually supplied by *lagging,* which is the process that uses *shrink tape* to supply pressure during cure. After the pre-preg wrap is completed, it is overwrapped with a shrink tape such as Tedlar. When exposed to heat the tape shrinks and applies pressure to the wrap. Some thermoplastic tapes show considerable shrinkage because they are prepared by stretching under heat followed by cooling while in the stretched position. In effect, they are thermoformed to new dimensions, but true to their thermoplastic memory, they revert back to their original shape when exposed to sufficient heat. Other methods for applying pressure to tube wraps include vacuum bagging, elastic tape, and the use of a female sleeve.

Elastic tape is applied by stretching it around the tube layup so that the pressure is immediate. Any stretchable material can be used but silicone rubber is the most practical because of its resistance to high temperatures and its no-stick properties.

A *female sleeve* is a rubber tube, usually of silicone rubber, that is prestretched, placed over the tube wrap, and then allowed to collapse to its original dimensions. Stretching is accomplished by inserting the sleeve inside a metal tube and pulling a vacuum so that the sleeve is pulled tight to the inside walls of the metal tube. To make the sleeve airtight, both ends must be forced around the ends of the metal tube. Valve location can be almost anywhere on the metal tube. If a vacuum valve is attached through the inside of the layup mandrel the rubber sleeve can also be used as a vacuum bag to add more pressure to the layup. Female sleeve are normally used for pre-preg–wrapped tubes that require both inner and outer smooth surfaces.

Spray Up

This is a process that uses a multiheaded gun to blast resin, catalyst, and chopped glass fibers simultaneously onto a mold surface. The resin normally used is polyester. After the desired thickness has been built up, the surface is rolled smooth to expel air and provide adequate density. The system is particularly suited to large sections because of the greater ease and speed of application compared to hand layup of glass cloth. However, layup with glass cloth provides better product uniformity and closer tolerances.

Continuous Pultrusion

Continuous glass roving is impregnated in a resin bath and drawn through a die that sets the shape of the stock and controls the resin content. A pulling device draws the shaped roving into an oven where it is given a final cure. The pultrusion process can be seen in Figure 11-1.

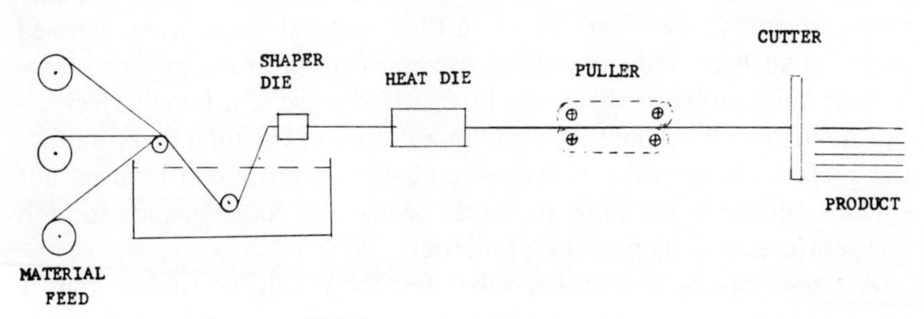

Figure 11-1. Pultrusion process.

Embedding with Polyester Resin

1. Mix enough resin to at least cover the entire bottom of the mold or to the desired thickness. Use 4 to 5 drops of MEKP catalyst (methyl ethyl ketone peroxide) per 4 tablespoons of resin.
2. Mix enough resin for the second layer about 15 min. after the first pouring. When the first layer becomes jellylike, dip the items to be embedded in the fresh mix and lay them on the jelled layer. Carefully pour in enough of the catalyzed mix to cover the embedment.
3. The third layer is usually pigmented. Mix enough resin to cover the second layer. Add the catalyst, pigment, and surface hardener and stir gently. Surface hardener is a mixture of a monomer, usually styrene, and wax. The wax rises to the surface, allowing quick cure of the surface area. Without surface hardener, the casting would have to be postcured in an oven to get rid of surface tack. Add the pigmented mix when the second layer has jelled.

Helpful Hints

1. Be sure to dip all embedments in catalyzed resin before placing in position. This is the secret in elimination of trapped air.
2. After you have mixed and poured the resin, use a needle or toothpick to break up surface bubbles, if any.
3. Polyethylene molds are best because you can remove the casting just by flexing the mold. You can also check for trapped air by holding the mold over your head and looking through the bottom. Even though poly molds are opaque, you can see the casting fairly well.
4. It is always best to stay on the short end of the catalyst range. This is a safe practice and produces the best castings. Don't rush. After addition of the last layer, the resin mass will exotherm, giving off considerable heat. Do not remove the casting from the mold until it is back to room temperature. If you add too much catalyst the poly mold will probably buckle from excessive heat.
5. Use more than the nominal amounts of catalyst on damp or rainy days and less on hot, dry days. After some experiences you will be

able to determine exactly how much catalyst is needed for each situation.

6. Never pour over a completely cured layer. Shrinkage of the cured resin creates a gap adjacent to the mold. Subsequent pourings can fill the gap with unwanted material.

ADHESIVE REPAIRS FOR SHOP AND HOME

Small voids are repaired by simply filling with parent material (resin mix used to make the part). Tiny openings are difficult to repair because the resin has a tendency to migrate out of the opening, especially if heated. For this type of repair, the openings must be made larger.

Polyester resin is used for repairing polyester parts but for metals, dissimilar surfaces, and when strong structural bonds are required, epoxy is best.

Rubber contact cements are used for repair of rubbers (except silicones) and for applications in which speed instead of high strength is the prime consideration. A repair with a contact cement can be made in 5 to 10 min. A coating is applied to each surface, then, when an aggressive tack develops, the adherends are joined together. Contact cements, as well as urethanes, are also used for repair of other flexible parts.

Aluminum-filled epoxy is used to fill holes in aluminum. When cured, the area can be drilled and tapped. This method can be used to relocate threaded holes.

One mil (nominally 0.001 in.) fiberglass is excellent for the reinforced repair of small or delicate articles, especially those having sharp angles and curves. When impregnated with medium-viscosity epoxy, it will conform and stay put whether horizontal or vertical.

Almost any broken part, except those with no-stick surfaces, can be repaired with epoxy adhesives or epoxy resin and fiberglass.

For general repairs around house and office, many good adhesives are available in hardware stores. Epoxies are the best all-purpose cements. Some of the acrylic-based adhesives, with almost immediate holding action, are excellent for bonding together small broken pieces. Eastman 910 is a product typical of this group.

For repairs involving reinforcement, remember fiberglass-epoxy, which is good for any repair where standard epoxy adhesives will not hold. The materials can be purchased at well-stocked hobby shops, and they are easy to use. Bond the broken ends together first, then either lay on or wrap around the wetted fiberglass. Remember to sand the bonding surfaces first. For cracks in housings, a piece of epoxy-wetted fiberglass placed over the crack is all that is needed.

Rubber contact cements are very handy to have around the house or office for repair of women's footware. When a sole becomes delaminated or a heel comes off or breaks in two, a little cement on each surface, a brief wait, and—presto!—the lady is back in business.

For large openings or gaps in fiberglass laminates, pieces of freshly impregnated fiberglass can be fitted inside the opening, then covered with several overlapping surface plies and more resin. Even ends that have been broken off can be rebuilt by using a fabrication aid having the same shape as the mating surface of the broken part. Repair as follows:

1. Apply a piece of fiberglass (1–2 mil) so that it covers a portion of the mating surface and extends across the area to be built up. (Do not extend beyond net dimension.)
2. Cover the fabrication aid with Teflon tape or release agent and mount it in place against the wetted fiberglass.
3. Lay on enough pieces of heavier fiberglass (with epoxy) to build up part thickness in the patch area.
4. Apply one to three pieces of fiberglass across the top of the joint.
5. Cure and sand as needed.

When making silicone-rubber–aluminum repairs, it must be remembered that the aluminum must be primed before bonding. It is not necessary to prime the rubber. There should be a 1-hour wait at room temperature between priming and bonding.

Most repairs are substantially stronger when the surfaces are sanded and solvent-cleaned before bonding.

Repair of Plastic Eyeglass Frames

This is a repair that demonstrates the holding power of thin fiberglass. The fixture used for repairs consists of two flat pieces of wood, one mounted perpendicular to the other and two clothespins mounted on the upright, two inches apart. Each pin is held by one screw, so that there is ample movement for adjustment. The repair is for frames broken in two at the nose piece.

1. Sand nose piece of both halves with a small Swiss file. (File in single strokes away from the lens.)
2. Dip broken ends in a shallow layer of acetone, shake off excess, mount in clothespins of fixture and hold in alignment for one or two minutes. If the ends are not bonded together at this point, dry them, apply a drop of contact cement and repeat the holding operation.
3. Coat broken joint with a mixture of epoxy and any quick curing amine such as DTA (diethylene triamine).
4. Apply a piece of 1 mil fiberglass and wrap it around the broken joint. Impregnate and add a second ply. A small artist's brush can be used for this operation. Once the fiberglass is in place, stop dabbing, or the glass strands will unravel.

With some frames, this type of patch is hardly noticeable. The bond will break only if flexed in a flexural bend.

Ear pieces broken away from the hinge can also be repaired by fiberglass and epoxy, without acetone or contact cement. Simply mount the pieces so that they will not move during cure of the patch.

In the absence of the materials listed, store-available epoxy and a dry cloth will provide a good temporary patch.

12 Plastics in Electronics

PLASTICS PROCESS LAB*

Most of the work performed in this lab consists of vacuum encapsulation, potting, foaming, adhesive bonding, and the coating of small circuit boards and other electronic components and assemblies. Equipment used includes vacuum encapsulators, lab-type vacuum-forming machines, ovens, pressure vessels, spray guns, heat guns, air guns, measuring devices, bell jars, drills, bandsaws, and hand tools.

Plastics technicians, of course, are key figures in these operation. Their knowledge and know-how are especially appreciated in a business where most workers are electronics-oriented but short on knowledge of plastic materials. A knowledgeable plastics technician in an electronics lab is in an excellent position for advancement.

Vacuum Encapsulation

Vacuum encapsulation is the process of covering an object under vacuum with a plastic resin mix so that when cured both the object and encapsulant will be free of entrapped air and vapors. This system is often used to encapsulate small electronic assemblies for use in outer space in order to protect them against shock, vibration, humidity, fungus, pressure change, and other damaging conditions.

The encapsulation sequence starts with the electronic unit being enclosed in a metal mold that has openings for material entry and wire passage. After the assembly has been placed in a container, it is positioned in the encapsulation machine, the vacuum is drawn, and the unit is evacuated. This is followed by introduction of the material through a funnel (usually glass) and into the mold. With the unit already under vacuum and the material being fed in under vacuum, a gradual removal

*Although this chapter was structured primarily for technicians working in aerospace electronics labs, much of the discussion applies to other labs within the plastics industry.

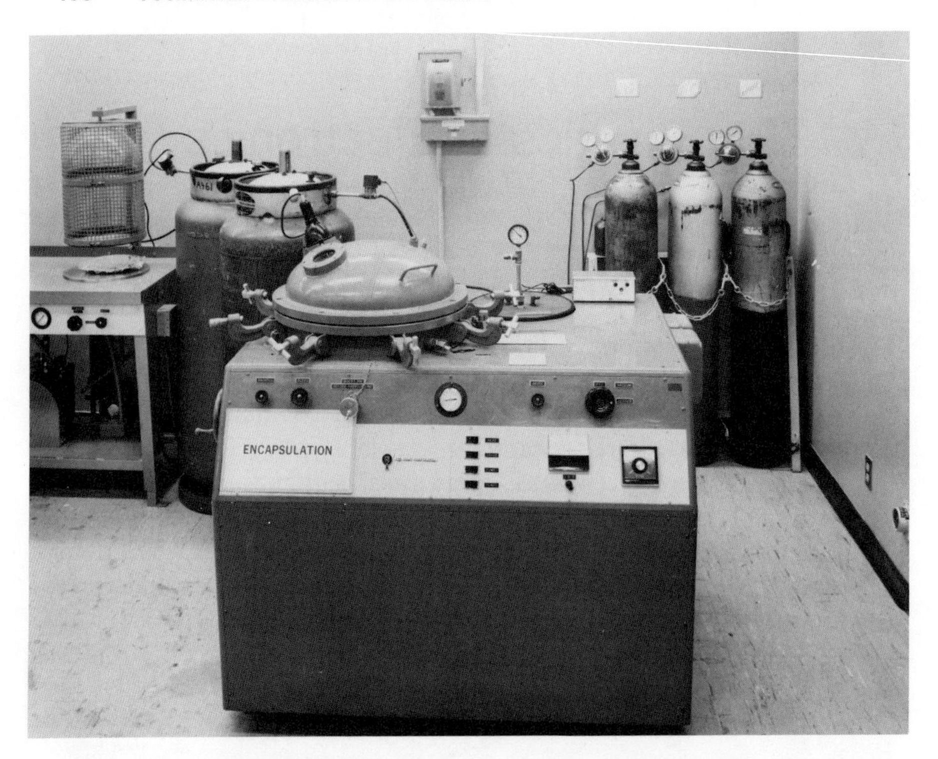

Figure 12-1. Vacuum encapsulating machines. At the left is a bell jar and 2 tanks of liquid nitrogen. To the right are 3 bottles of dry nitrogen gas. *(Courtesy of Hughes Aircraft)*

of all air and vapor bubbles is effected. The flow of material is controlled by a stopcock located in the pour funnel. At the conclusion of the vacuum phase the resin is cured under heat and pressure. Encapsulating machines (Figure 12-1) are available with facilities for vacuum, pressure, and heat, so that the unit need not be removed until the entire operation has been completed.

Encapsulation Sequence

1. Assemble the mold around the unit to be encapsulated, being careful to thread the wires, if any, through the proper mold openings. Where there are several wires going through the same opening, they should be fed through one at a time to prevent twisting.

2. Place the mold in a poly container that is slightly larger than the mold at the sizes and 1–2 in. taller at the top. If there are bare wire leads protruding through the top plate they can be protected by either: (a) lightly greasing and covering with Teflon tubing or (b) coating with an appropriate RTV.
3. Install the pour funnel in the lid, place the assembly on the shelf of the vacuum chamber, secure the lid, and introduce the vacuum.
4. Weigh and mix the material components and degas in a bell jar for about 5 min. Partial degassing allows the resin to be fed into the mold without the initial severe bubbling. It also helps to prevent a significant drop off in vacuum pressure during the feed-in operation.
5. Feed the resin in slowly enough to stay within the specified vacuum pressure limit. Follow the specified procedure, keeping in mind the pot life of the resin. All operations up to the start of curing must be completed within the allowable time.
6. At the conclusion of the vacuum cycle, turn off all vacuum valves and gauges and introduce the pressure. Pressure is supplied by an inert gas such as dry nitrogen or carbon dioxide.
7. Turn on the heat and cure per the specified time and temperature.

Helpful Hints when Vacuum Encapsulating

1. Molds for repeated use are normally made of steel. New molds must be solvent-cleaned then flushed with an air gun to remove tiny fragments that may be trapped in the threaded holes. New screws must also be solvent-cleaned. All mold surfaces must be treated with a release agent, preferrably silicone, which requires baking for approximately 1 hour at 150–200°F.
2. When working with epoxies it is best to fasten the mold with socket type head screws. They are easier to remove than other types. Before inserting, they must be coated with high vacuum silicone grease. A failure to grease the threads could result in twisting off of the screw heads during disassembly.
3. If the unit has protruding side wires, pot the area between the mold and the container with an RTV rubber such as RTV 511.*

*RTV 511 is a trademark of General Electric Co.

Figure 12-2. Lab thermoforming machine. Section at the top houses heaters. The levers are pushed down to force the work close to the heaters. Thermoforming sheet is polypropylene. *(Courtesy of Hughes Aircraft)*

Pot to just above the wires. This will prevent entrapment of the wires in the hardened encapsulant. The RTV will not adhere and is easily removed by hand. Unless the wires are protected, they can be easily damaged. A fast-acting curing agent such as Nuocure 28* can be used to solidify the RTV in 3–4 min.

4. Long, protruding wires can be coiled and tied and left to hang outside the container. The bare ends can be sealed with RTV.
5. Polypropylene is ideal for making containers. It is easily formed on a laboratory vacuum-forming machine and can withstand temperatures in excess of 280°F. Figure 12-2.
6. If you have trouble filling out the mold because the openings are too small or because the material is too thick, heating the mold

*Nuocure 28 is a trademark of General Electric Co.

and the material will thin out the resin and allow it to flow better. To include this operation, however, you must know the temperature limitations of the encapsulating material. Some materials can be heated to 250° F, whereas others cannot be heated at all because of a short pot life.

7. When encapsulating with thick, viscous materials, pour to at least ¾ in. above the top of the mold. Also, at final evacuation, introduce the air very slowly—barely cracking the valve—for at least the first 5 in. of mercury drop-off. This will prevent pockets from forming over the mold pour holes. When these pockets form during evacuation, you are apt to trap some air inside the mold.
8. When using low-viscosity (thin) mixes, it is advisable to place a rubber band around the funnel stopcock to hold it firmly in place. Thin mixes can seep through and break the vacuum seal.
9. The stopcock must be removed from the funnel and cleaned after each use. Even an undetectable amount of epoxy can "freeze" it up.
10. When disassembling, care must be exercised in removing cured resin from the mold. Where there is just a rubbery cap above RTV, pry the cap off the mold and pull the wires from the cap, one at a time. With brittle materials such as unfilled and unmodified epoxies, the shell can be removed with any chisellike tool and a plastic mallet. The pieces will fly, so care must be used to protect yourself and those around you.
11. Paper outgases under vacuum. Use plastic or metal containers when degassing.

Vacuum Impregnation

Vacuum impregnation is a process of coating under vacuum. In electronics it is used mostly for impregnating small, individual electronic components such as transformers and capacitors. The part is wrapped tightly with a reinforcement such as fiberglass (usually one ply) and held together with polyester web tape with tack on one side. Terminals and wires are covered with rubber maskant or RTV, following which the part is completely encased in polypropylene or other suitable thermoforming material. One bag is formed over the part, and then a second bag is formed in the opposite direction. Small slits are made in the casing with a pointed Xacto blade or similar tool. Obviously, much care must be used when performing this operation to avoid cutting wires. The best way to avoid damaging the part is to preselect slit areas before adding the thermoformed casing. The slits are made to allow penetration of the resin, which is usually a rigid epoxy. After the casing has been slit, the part is placed in a poly container that is just loose at the sides and 1 in. or higher at the top. After the vacuum cycle has been concluded the impregnation is usually cured under pressure at 50–300 psi. For small transformers, it is a good idea to pot the terminal area above the bag with RTV. This will afford added protection during disassembly.

It is surprising how a thin sheet of polypropylene can be formed, even over upright wires. The conformance takes place with no movement of the wires.

Potting

1. When open potting of only one area of a circuit board must be accomplished without tooling, you can use thin plates such as aluminum or glass-epoxy to contain the resin mix. Cut the plates to the desired height and length and tape them together leaving room at the joints for bending. Apply silicone release agent to the inner surfaces, and place the assembly over the area to be potted. Seal at the base and at the outer corners with RTV. Teflon tape can be applied on the inner surfaces in place of the release agent. If the plate assembly does not fit flush on the board, notch the areas to the height of the protrusions and use more tape and RTV.
2. When potting the backside of a connector with a thin, low-viscosity resin, there is always a possibility of the resin seeping through and onto the pins, if the pins are not tightly sealed. To prevent seepage, the cognizant engineer can specify one of the following:
   - A. Apply and cure a sealing coat of thicker material before potting.
   - B. Protect the pin area with removable RTV.
   - C. Allow the potting to get at room temperature before applying heat.
3. When a potted or encapsulated unit is being repaired, it must be thoroughly evacuated before patching if you want it to be air-free. Only in cases where shallow voids are clear of components

can you effect air-free patching by degassing the material only. If an oven is used for cure, it is wise to check for air bubbles during the first 10 minutes of cure.

4. When potting a section of an assembly having several attached parts, be sure that all of the parts can tolerate the cure temperature. If one or more sections cannot, you will have to localize the heat by using an infrared lamp on the potted area. Cover the adjacent areas with aluminum foil to protect them from heat.
5. Units too large for available oven space can be cured under an aluminum-foil tent. Lay foil down and place the unit on it. Position at least two infrared lamps to provide adequate heat, then tape foil sections together and hang them over the unit and lamps. With foil being so light, you should have no trouble finding something to support the tent. Use a temperature controller to monitor the heat, and locate the thermocouples close to the unit. Heat loss can be minimized by locating a sheet of foam or other nonconductor between the table and the unit. For small units, several pieces from an orange stick or wooden tongue blade located between the foil-lined table and the unit should suffice.
6. Pottings or encapsulations in an open mold can be cured under gas pressure without runoff because the pressure within the chamber is equalized in all directions.
7. When potting is cured in an oven, a small sample of the material should be cured along with it. This will tell you when cure is complete without having to touch the potting. The surface of the potting can be permanently distorted if touched during gelation.
8. Use an air-actuated dispensing gun when filling molds having large volumes or when filling several molds at one time. This is particularly effective when potting with materials that are thick and slow pouring. Injecting with 10–20-cc poly syringes can sometimes be very time consuming and even detrimental when working with short-pot-life materials. The dispensing apparatus consists of a cylindrical poly tube having a necked, down-threaded end and a fully open end. After a nozzle having the appropriate size and shape is screwed on, the tube is ready for use as soon as it is loaded with the potting material and mounted in the air gun.
9. Thermolite 12* can be measured by a medicine dropper because

*A trademark of General Electric Co.

of the small proportional amount used. Fifty drops of Thermolite 12 equal 1 gram.

10. In many instances, spillage or overruns of catalyzed red RTV is much easier to clean after curing. When cured, it parts easily from dissimilar surfaces. When wet, it spreads and smudges.
11. The curing of clear, water-white RTV is sometimes inhibited when it is in contact with the adhesive on some adhesive tapes.

Curing at elevated temperatures (150–160°F) rather than at room temperature seems to alleviate this sensitivity.

Foaming with Rigid Urethanes

Rigid urethane foams are sometimes used to encapsulate small electronic assemblies to provide them with a measure of protection in outer space. One advantage of foam encapsulation is the ease with which the cured foam can be removed. It is sometimes necessary to replace a malfunctioning component. When this occurs the foam can be easily removed with a pointed orange stick.

The most commonly used foams are known as "2-lb" and "4-lb" foams. Actually, they are foams weighing 2 lb and 4 lb per cu ft. For an idea of how lightweight these materials are, compare them to water, which has a density of 62.4 lb per cu ft.

Small circuit boards are usually foamed in aluminum molds. The board is encased in the mold, and the mixed liquid is injected through holes in the top plate. The holes also provide for escape of excess foam. When the liquid starts foaming (called *blowing*), the pressure buildup is intense, and adequate escape must be provided.

Components mounted on a metal chassis with only an upright at each end can be foamed by clamping two smooth, released aluminum plates against the two open sides and applying masking tape loosely across the top after the mixed liquid has been poured in. The tape provides the measure of restraint that is necessary to achieve a void-free encapsulation. This type of chassis is known as a *wrap around*.

Assemblies having four sides and a base (usually aluminum) are referred to as *tubs*. After the mixed liquid is poured in a released aluminum plate having a sufficient number of holes is clamped on across the top. Assemblies containing delicate wires or components are sometimes foamed completely without restraint to eliminate back pressure during foaming. This is known as *free blow* or *open blowing*. When this system is used it is important to wet out all of the components when pouring. Failure to do so will result in large voids directly above the components. Where possible, it is best to use a cover with many large holes so that back pressure will be minimal. Delicate components would still be safe, but voids and loss of density and strength would be minimized.

Rigid foam materials are provided in two component kits. Because of the short time between mixing and blowing, all preparations should be completed before the start of operations. After foaming, the unit should be allowed to stand for 30 min at room temperature before oven-curing. Premature curing can result in excessive blowing. A small amount of foam expansion will take place in the oven, even when the 30-min wait is observed. Slower-blowing foams are usually poured into molds that have been brought up to 110° F. This assures good distribution and uniform density.

Helpful Hints when Foaming

1. Urethane foams are toxic and should be mixed and poured under a hood.
2. The first consideration when foaming is where you *don't want* the foam. All areas that must be kept free of foam can be protected by covering with masking tape. Release agents can also be used, but some scraping will still be necessary. Release agents should not be used on surfaces that will be subsequently adhesive bonded.
3. Before foaming it is wise to check for loose or damaged wires and components. Also, a check should be made to be sure that there are no components or wires protruding too high. When in doubt, pass a metal straightedge across the open top of the tub or mold. If contact is made with the straightedge, it should be brought to the attention of the supervisor.
4. Whenever possible, it is best to pour into an open top then secure the top plate—instead of injecting the material through the top plate. This will eliminate the possibility of incomplete filling out of the mold. When injection is mandatory, it is a good practice to block off several holes for several seconds to force the foam into the corners. This is particularly true of some 4-lb foams that do not have the strong blowing action of 2-lb foams.

5. Threaded holes on the top surfaces of tubs or other assemblies can be protected by covering them with thin (3–4 mil) Teflon adhesive tape. The surfaces that are narrow side and/or partition edges should be cleaned with acetone or MEKP before applying the tape. This will insure good adhesion of the narrow strips. Trimming can be accomplished with a sharp, pointed Xacto blade or similar tool. To avoid deep penetration and possible damage to components, only the tip of the blade must be used.
6. New molds should be: (a) solvent-cleaned, (b) coated with wax, and (c) coated with a fluorocarbon. For subsequent encapsulations, the application of a fluorocarbon will suffice.
7. Four-sided units having thin sidewalls with no reinforcing ribs must be foamed in a holding fixture to prevent bowing. In the absence of tooling, the walls can be protected by clamping on metal plates. The clamps should be just snug enough to prevent wall movement.
8. Some units have recessed tuning screws that must be kept free of foam. Appropriate-size Teflon tubes placed over the screws will give ample protection. Teflon rods can also be used. Emplacement should be gentle. Undue force can rupture a delicate solder joint. To remove a tube or rod, first turn it clockwise or counterclockwise a half-turn to break the seal before pulling it out.
9. Four-lb, rather than 2-lb, foam should be specified for shallow two-sided encapsulations of small circuit boards. Two-lb. foam, which is used for most electronic encapsulations, is sometimes too delicate for this application. The problem lies in keeping the foam intact during disassembly.
10. Vendor containers must be carefully checked before weighing, so that the A-part from one kit is not used with the B-part of a different density kit. Soft or rubbery oven-cured rigid urethane foam usually indicates improper weighing or a mixed kit. Powdery foam usually indicates undercure. When undercured, the foam can still be cured properly if exposed to the appropriate heat cycle.
11. Pour time can be shortened when working with quick-foaming viscous materials by using a paper cup to weigh the materials in and then cutting the cup down to just above the liquid level before mixing.

ADHESIVE BONDING

Adhesive bonds provide wider distribution of stresses than do mechanical fasteners. In addition. bonded structures are usually lighter in weight, lower in cost, and easier to assemble. They can also be tailored to provide specific characteristics to the bonded joint.

Proper preparation of the surfaces to be joined is very important in adhesive bonding. In most cases, smooth, glossy, or contaminated surfaces provide poor bonds with chemically reactive adhesives.

As the first step in any surface preparation, both metallic and nonmetallic surfaces should be degreased with a solvent. Contaminants can be pushed farther into the adherend by sanding. They can also inhibit chemical etchants from performing with maximum efficiency.

Most plastic surfaces are usually just sanded and solvent-cleaned again to make them bondable, though nonstick surfaces such as Teflon and polyethylene do require special preparation. Thin, unreinforced thermoplastics should be sanded with a fine grit in order not to create weak spots, whereas reinforced thermosets can be sanded with coarser grits. Sanding the surface creates peaks and valleys which, in effect, provide a larger bonding area.

MEK, acetone, and similar solvents will attack thermoplastics such as cellulosics, acrylics, and polystyrene. When in doubt about the appropriate solvent, use isopryl alcohol for cleaning.

Epoxy Adhesives

Much of the bonding performed in an electronics plastic lab involves the use of epoxy adhesives. They are quick-curing structural materials featuring good adhesion to dissimilar surfaces and good resistance to moisture, chemicals, and solvents. Most contain fillers that impart specific characteristics to the bonded joint. Silver powder is used for electrical conduction, aluminum oxide ensures thermal conduction, and silica powder serves as a thickening agent. Several other fillers are used—each imparting a specific property to the adhesive.

Curing Agents

Room-temperature-curing agents like triethylene tetramine (TETA) are used extensively in epoxy adhesives. Also used abundantly are polyamides such as Versamid 140. Versamids can be used alone as hardeners or mixed with other curing agents and function as modifiers. In addition to acting as curing agents, polyamides also serve as softening agents for epoxies. They reduce brittleness and add peel strength and impact strength to the bonded joint. Versamids have an unmistakably sweet, nutty smell and can be easily detected in epoxy formulations. Versamid-epoxies will cure at low or ambient temperatures.

Miscellaneous Adhesives. Other adhesives used in electronic assemblies include polysulfides, silicones, and elastomeric contact cements. Polysulfides are synthetic rubbers used to fillet and bond components on circuit boards, to protect them against shock and vibration.

Silicone Rubbers and Adhesives. Silicone (RTV) rubbers are used to bond silicone rubber to rigid substrates—usually aluminum. The aluminum must be primed about an hour before bonding: it is not necessary to prime the rubber. Most RTV rubbers will not bond to dissimilar surfaces unless they are primed. RTV adhesives, however, will adhere to dissimilar surfaces without the use of a primer, though primers are usually recommended to improve the bond strength.

Contact Cements. Contact cements are used primarily to bond aluminum to rubber substrates (except silicones). A coating is applied to each adherend, and when an aggressive tack develops (in several minutes), they are joined together. Once contact has been made, the adherends are difficult to move. Correct alignment should be made at first contact.

Premixed Adhesives. Many electronics companies purchase premixed adhesives in a frozen condition. They are packaged in small plastic syringes and can be thawed out easily by holding them in the palm of the hand for 2–3 min. Some companies prefer to make their own adhesives as the need arises, and some use a combination of both systems. Premixed adhesives are normally stored below −40°F for a maximum of 30 days.

Five-mil-diameter microspheres are sometimes added to adhesives to insure a 5-mil glue line. They are normally used in the proportional amount of 1% of the weight of resin. Because glass spheres are hygroscopic (they pick up moisture), they should be kept in dry storage.

Bonding Techniques

1. Most adhesives are applied to all faying (bonding) surfaces. This is especially important when using highly filled adhesives, when wetting may be a problem. If wetting is a problem because of high filler content and/or rough surfaces, a thin preliminary coating of catalyzed parent resin to the bare surfaces can be very helpful.
2. Smooth, glossy surfaces do not provide strong bonds.
3. When buttering the surface of a small, flat, rigid adherend, use masking tape on the underside to hold it so that the adhesive will not come in contact with your hand. Solvent-cleaning the surface first will make the tape hold better.
4. When ready to mix adhesives for cold storage, make out the identification tags and have them inspection-stamped before starting. Most of the adhesives used in electronics have a short pot life, which helps prevent advancment of the adhesive before completion of the operation. Care should be exercised when filling syringes in order not to leave any on the outside surfaces. A dirty syringe is not only irritating to the user but also can expose the user to dermatitis.
5. When packaging adhesive components for on-site mixing, use the "wet bottle" method if the curing agent is a low-porportion type such as dibutyl tin dilaurate (0.5–1% in RTVs). Transfer some of the curing agent to the container, then pour back out as much as you can. Add the exact amount needed on top of what is left in the container.
6. When applying adhesive to a hard-to-reach area, you may be able to gain access more easily by connecting a Teflon tube to the tip of the syringe. Tape it on securely with Teflon tape.
7. Double-back tape can be used to bond small, flat lightweight objects on horizontal or vertical surfaces.
8. Be alert to the existence of different cure times for the same adhesive. Polysulfides, in particular, come with a long pot life or a short pot life. The adhesive with a long pot life is designed for large bonding operations requiring more working time. The manufacturer usually indicates the type on the label with just one or two letters or numerals next to the name of the adhesive.
9. On occasion, the need may arise to remove a bonded component

from an assembly. For rigid bonds, this is best accomplished by saturating the component and substrate with a cold-inducing spray and applying one of several methods for removal. Using a thin chisel-blade scraper, tap the blade into the joint. If the component is in a metal housing with no tool access, slightly flex the housing base or tap the outside gently with a rubber mallet. Needless to say, much care must be exercised when performing this operation. The situation will dictate the extent of caution necessary. For removal of cured pre-preg or thick coating from steel surfaces, heat to 300–400°F and scrape off the adhesive with a chisel blade while hot. Be sure that the unit can tolerate the high temperature before attempting this operation. Some materials will scrape off at 200°F.

10. When time permits, it is advisable to gel room-temperature-curing adhesives at room temperature rather than at elevated temperatures. This applies primarily to bonded assemblies whose ability to remain intact when heated is uncertain. When gelled, usually within 2–4 hours, the assembly can then be safely placed in an oven for further curing.
11. Small sandbags can be helpful in supplying contact pressure for multicontoured or irregular adherends.
12. When mixing thick, heavily filled resins with water-thin curing agents, stirring must begin very slowly. Rough or hurried stirring almost always results in splashing out part of the curing agent.
13. To extend the pot life of quick-curing epoxies, transfer the mixed adhesive to a clean aluminum plate and spread it out. The larger the surface area of the adhesive, the longer the pot life.
14. When working with thick "gunk," secure the container to the table with double-back tape. This leaves you both hands free instead of using one hand to hold the container. If the base of the container is recessed, use masking tape.

Surface Preparation

Aluminum surfaces are usually treated with chromic acid etchant before bonding. These is not a great difference in lap shear strength between aluminum specimens prepared by chromic-acid etching and those prepared by sanding, but there is some question as to the ability of sanded aluminum joints to retain bond integrity upon aging.

Degreasing with Detergents. Some fabricators believe that some solvents can contaminate a bonding surface. For this reason they prefer to clean adherends with a detergent solution. The adherends are immersed in hot water containing any commercial dishwashing detergent. They are then rinsed several times with fresh water.

Prepared adherends should be assembled as soon after treating as possible to avoid recontamination. The surfaces can be protected by wrapping or covering with polyethylene or other suitable film. When long delays are anticipated the faying (bonding) surface can be protected by application of a primer. Prepared surface should not be touched before bonding, especially with vinyl gloves, which can exude plasticizer. The new-car odor inside automobiles is primarily due to the oil/base plasticizers used to soften vinyl upholstery. The drying and exuding of the plasticizer is responsible for cracks in upholstery sometime found on the shelf above the dashboard in older automobiles.

	Parts by Weight
Water	30
Sulfuric acid, 66° Baumé (Bé)	10
Sodium dichromate	1–4

Etching of Aluminum

Submerge the degreased specimens in the above bath at 150–155°F for 8–10 min. Rinse well in tap water, then rinse in distilled water. When the specimen is rinsed, there should be a continuous unbroken sheet of water covering it. A break in the water layer indicates incomplete cleaning and a need for rework.

When making the chromic-acid solution, several precautions must be observed. Never pour water into acid. Start out by transferring a measured amount of water to the mix container. Next, add the acid *slowly,* then add the sodium dichromate. Mixing should be done under a hood to avoid breathing of the vapors.

After they are cleaned, the specimens should be dried in an oven at no higher than 150°F. To avoid water marks, before oven-drying the specimens can be purged quickly of water with a heat gun.

Aluminum sections that cannot be immersed in the acid solution because of their size, location, and so forth can be treated as follows:

	Parts by Weight
Water	80
Sulfuric acid, 66° Bé	55
Sodium dichromate	10
Fumed silica powder	enough to make a thixotropic paste

1. Scrub or solvent clean to obtain water-break-free surfaces.
2. Apply the paste and let stand for 25 min. Reapply as needed to prevent the mixture from drying out or turning green.
3. Remove with tap water and do a final rinse with distilled water.
4. Dry with a heat gun or other means at temperatures to 150°F.

Etching of Steel

The following process is one of several used to etch steel, preparatory to the bonding operation.

	Parts by Weight
Water	80
Sulfuric acid	10
Oxalic acid	10

1. Immerse in the acid solution for 10 min at 180°F.
2. Scrub until clean under running tap water.
3. Rinse with distilled water. Check for water break.
4. Dry at 150°F.

Sodium Etch for Teflon

	Composition
Sodium metal chips	45 grams
Naphthalene	125 grams
Tetrahydrofuran	1 liter

1. Transfer the tetrahydrofuran to a polyethylene or stainless-steel container with a snap-on polyethylene lid.
2. Cut the sodium into small chips and immerse immediately in the tetrahydrofuran.
3. Add the naphthalene.
4. Stir until the sodium and naphthalene have completely dissolved. At this point, the solution should be brown in color. The lid should be taped on when the solution is not in use to prevent the escape of gases and intrusion of air.
5. Immerse the Teflon in the solution at room temperature for 5–20 sec.
6. Rinse the specimen with solvent or distilled water. If properly etched, the Teflon will be colored light brown to dark brown.

Precautions

1. Work under a hood and wear rubber gloves when handling sodium. It reacts with air or moisture to form a strong caustic compound.
2. Avoid direct contact of sodium with water, chlorinated hydrocarbons and dry ice. Prepared sodium etchants can be purchased from W. L. Gore and Assoc., Newark, Delaware (Tetra Etch) and W. S. Shaban, Newbury, California (Bondaid).

COATINGS

Many electronic circuit boards used in aircraft and spacecraft contain some type of protective coating. Urethane and epoxy coatings are among those used extensively. Epoxy coatings are moisture and chemical resistant. Some epoxies are made tougher by addition of a polyamide or other softening agent. Urethane coatings are tough, strong, and mar resistant. Some contain a mineral filler but, in general, most coatings for electronic circuit boards are transparent and unfilled.

Many components and housings for spacecraft are coated with paints that impart a specific characteristic to the surface. Some silver-plated components are so delicate that just touching them with bare hands can inflict a stain. They must be handled with white gloves and wrapped in "silver paper" when in storage.

Coating Tips

1. Circuit boards are usually cleaned with Freon or isopropyl alcohol before being coated. Acetone or MEK should never be used as solvents because they attack certain thermoplastic materials that may be part of the board assembly.
2. The adhesive from some tapes will stain silver plate when heated. Be sure to use only specified tapes when masking.
3. When spraying circuit boards having upright components, you should spray from five directions. To insure complete coverage, make one pass at an angle from each of the four sides, then a final pass aiming straight down. Be sure you have good vision of the spray pattern. If you don't, hold the board up by the side edges in a position enabling you to see clearly.
4. Units coated with silicones should be cured in a separate oven designated "for silicone only." Vapors from silicones will inhibit the cure of some materials and leave no-stick areas on others.
5. After mixing a mineral-filled coating, it is a good idea to pass it through a paint strainer before using in order to remove lumps.
6. When painting a housing, use clean white gloves instead of all-purpose rubber gloves. Handling any silicone with the rubber gloves probably leaves no-stick areas on wherever they are touched.
7. A common practice when oven-curing small parts is to place them on an aluminum-foil-covered rack to prevent their falling between the rack bars. When this is done, small holes or cuts should be made in the foil in order not to restrict the normal circulation of hot air.
8. Unused circuit board terminals are normally covered with a rubber-latex maskant before coating to keep them resin free. When applying the maskant, use enough so that it can be easily removed after the coating operation. If you apply it thinly, it will come off in pieces.
9. Most coating materials contain solvents; therefore, boards coated with these materials should be left at room temperature for at least 30 min before curing to evaporate most of the solvent. It is also a good practice to cure for at least 10 min at 120–130°F. before applying the specified temperatures, which is usually 150–160°F. Most coatings will blister if exposed to heat too soon.

10. Some urethane coating materials are very sensitive to moisture. The pouring area of cans containing these materials must be thoroughly cleaned each time after using. If material is left around the cap or lid area it will froth and harden, making it difficult to get the closure on and off. It will also provide access for moisture to get inside the can and drastically reduce the shelf life. For containers with screw-on caps, clean the pouring surfaces, and place a piece of polyethylene film over the opening before screwing on the cap. This will keep the threads free of foam. A section of a poly bag will serve the purpose.
11. When spraying, keep the gun in motion at all times. A hesitation can result in a runny, nonuniform surface. To keep it functioning properly, the spray gun must be thoroughly cleaned after each use.
12. Strippers are available to remove epoxy paints and coatings from metal surfaces. The stripper does not harm the metal but attacks most plastic surfaces.
13. Boards coated on both sides can be dried and cured with the aid of paper clips bent and shaped into "S" hooks. There are usually holes in the four corners of a board so that the board can be easily suspended by paper-clip hooks from the oven rack. On rare occasions when there are no corner holes, the board can be supported on four cut-down orange sticks that have been positioned upright on a cured-foam slab. The ends of the orange sticks are sharpened for easy penetration into the foam.

EPOXY PAINT STRIPPERS

These are chemical preparations that remove unwanted epoxy paint or coatings from metal surfaces. Some are also effective for stripping urethanes. Most epoxy strippers contain phenolic and chlorinated solvents, so all necessary precautions should be observed when working with them. Rubber gloves should be worn, and work should be conducted under a hood or in a well-ventilated area.

After the stripper has been applied, a dwell time of approximately 5 to 15 min is necessary for thorough loosening of the coating. When the coating is loose, all areas are thoroughly agitated with a stiff bristle brush. When all the film has been removed, the surface is rinsed under cold water.

Epoxy strippers attack most plastics; therefore, plastic areas, as well as all rubber areas, must be protected during the stripping operation. When only part of the paint is to be removed, the area to be protected can be covered with masking tape. MEK or acetone will strip many nonepoxy coatings.

FABRICATION AIDS

Nonproduction items such as wooden tongue blades, cotton-tipped applicators, clamps. paper cups, masking tape, sandpaper and double-back tape can be used to advantage by the alert technician or fabricator. RTV rubbers are also very valuable fabrication aids.

In addition to their value as stirring rods, wooden blades can also be used as shims, scrapers on delicate surfaces (sharpen edge), squeegees (wrapped in Teflon tape), protection for clamped surfaces (tape the blade in place), and as measuring devices for cutting or marking.

Cotton-tipped applicators can be used to clean sharp corners by flattening the cotton tip with smooth-jaw pliers. Sharpened to a point, the wooden end can be used as a tool for applying adhesive to tiny bonding areas and for applying rubber maskant to terminals on circuit boards.

Masking tape can be used in many different ways for holding or securing operations. It can also be marked and used as a measuring guide, sign, or label. The sticky surface picks up small, hard-to-grasp foreign matter from parts, work surfaces, or clothing.

Double-back tape, with adhesive on both sides, it useful in a large variety of holding operations. For temporary holding, only a few small pieces should be used, or difficulty will be encountered when the bond is disassembled.

In addition to the standard applications of clamps, they can hold together assemblies and fixtures. In the absence of a tool for foaming, potting, or bonding, think of clamps and plates. It is surprising how many "hot" (urgent) jobs can be finished quickly without immediately available tooling through the use of clamps, plates, and ingenuity. A soft rivet that cannot be reached with a conventional tool can be easily fastened by holding the rivet in place with masking tape, standing a metal plate against it, and squeezing together with a C-clamp.

Paper cups, when inverted, can serve as small work stands and holding fixtures. Cutouts can be made in the base to accommodate protrusions on small parts, thus facilitating holding and curing operations.

Wide-mouth cups, for example, those used to package cottage cheese, are particularly useful around the lab. Small parts for bonding or potting can be secured on the inverted cup base with double-back tape. This again will make work and cure easier. Several inverted cups can be used as a stand on which to place large parts.

Sandpaper can be used to make surfaces skid proof. To prevent a nut or other threaded component from slipping, simply tape two pieces of sandpaper together with double-back tape and insert in the slide area.

RTV rubbers can aid production in almost any plastics lab or shop. These materials do not stick to any unprimed surface (except other silicones) and can be cured in 2 min with Nuocure 28. They can be poured, or inserted as solids into any area that is to be kept free of foam, potting material, adhesive, etc. When an area to be checked for dimensions is in a restricted or hard-to-reach location, the task can be made easier by casting RTV into the area. Dimensions can then be taken from the cured RTV. Powdered silica can be added to the resin to prevent runoff.

RTVs can also be used as sealants when spot-potting or foaming areas on circuit boards or other assemblies. With this system, glass-epoxy or other flat pieces are taped together around the area to be filled. RTV is then applied to the outside corners and at the base joints to insure a complete seal around the enclosure. White RTVs such as RTV-11 serve well for this purpose. Solid (cured) pieces of RTV come in handy for wedges and for placing over sections being bonded so that they can be held in place by masking tape. Some components within an assembly are difficult to tape down because of higher adjoining components, but a large slug of solid RTV over the top adherend will allow the tape to be secured around the outside walls of the assembly, if necessary. RTV-11 is easily removed when cured because of its relatively low-tear strength and no-stick properties.

An acid brush cut down to approximately ⅛ in. serves as an excellent scrubber in areas requiring a stiff bristle. It is very helpful in removing resin buildup on files (use with acetone).

Learning to use standard supplies and equipment as fabrication aids is a great asset to a technician. It not only helps speed up work, but also makes the job easier and the satisfaction greater.

When presented with what at first appears to be an insurmountable work problem, do not be dismayed; think of fabrication aids. In a recent observation, a bright technician was asked to screw a self-tapping aluminum screw into a hole that was located at the base of an opening 7 in. long with a diametter of ½ in. The technician first tried to insert the screw with a screw-holding screwdriver. This did not work because the holding spring could not release the screw after it was inserted: the opening did not provide enough room for separation. Undaunted, the technician removed the screw, taped it to the end of the screwdriver with masking tape, and reinserted it. After the screw was emplaced, the technician squirted a few drops of MEK onto the tape to break the bond and free the screwdriver. The technician was asked if there was anything that could have been done had the tape not worked. The answer was "Sure, I would have spot-bonded the screw to the screwdriver with a quick-curing adhesive. This would hold the screw securely and still allow me to break the bond after the screw was inserted." The moral of this story is "There is always a way." All you have to do is find it.

GOOD WORK HABITS IN THE LAB

1. Keep your work area neat and uncluttered. If brown kraft paper is available, use it to work on and change it as needed.
2. When you are through using a balance or other equipment, be sure to check for spillage each time. You will be amazed how often you will find at least one or two drops somewhere. Work neatly. If you do make a mess, clean it as soon as possible. Keep in mind that clean work habits reflect good organization.
3. Organize your tools so that you don't have to fumble around when you need one.
4. When you receive an assignment, study the print and process instructions thoroughly. Don't just look for the main directives on a print. Read all of the statements. If you don't, you will miss something sooner or later. Always look for engineering changes (e.o.s), which are changes made after the print has been released.
5. Be sure that previous operations have been "bought off" (approval stamped) by inspection before starting an operation. This is done on the process worksheet.
6. When you receive an assignment that you have had before, never proceed without first looking at the print and paper work. Something may have been added or deleted since you last worked the job.
7. When you work with a process spex, don't just check for mate-

rials and proportions. Look up the pot life and all the alternate cure cycles. This will eventually enable you to answer such questions as "Can it be cured at room temperature?" "How can we shorten the cure cycle?" "What is the cure cycle?" "How long do we have to work with the material (pot life?)" With proper interest you will be surprised how easily you can remember the details of so many different specifications. Keep a condensed spex list handy—one that lists the number of the spex and the main ingredient. If you ever do forget momentarily, a quick glance is all you need.

8. Keep a phone list handy—one that contains the numbers of all the people you deal with, such as engineers, planners and supervisors.
9. Be aware of all the work that goes on around you. Get all the information you can about every job being worked in your area.
10. When you receive an assignment involving two or more similar parts, the first thing you should do is check to see if they have been marked with a serial number. If they have not been marked, attach temporary identification to them. A piece of masking tape with the serial number written on it is sufficient.
11. When working with very small screws, washers, or other small hardware, stay close to the center of your table. This will prevent the hardware from falling to the floor, possibly necessitating a long search—and, in some cases, a replacement. Working away from the edge of the work table is especially important when handling any electronic component or assembly. They are delicate and could be easily damaged if dropped. Be especially careful when stripping masking tape. The pulling, jerking motion can cause a problem unless the unit is held firmly.
12. When you are through with a spex, return it back to the file. Return everything to storage that is not being used. This will keep your lab neat looking and more enjoyable to work in. Others will follow your good example.
13. Never raise the temperature of an oven without first examining the chamber to be sure it is empty. Heat sensitive parts can be damaged or destroyed if subjected to excessive temperatures. Also, when you do increase the temperature of an oven that normally operates at a set temperature, be sure to tape a large sign on the oven door indicating the new temperature. An alert tech-

nician never places a part in an oven without first checking the recorder for a true temperature reading.

14. When weighing several materials of a formulation, be sure to replace the lid or stopper immediately following each weighing. This will preclude the possibility of returning a material to the wrong container.
15. Small and tiny screws are sometimes difficult to get started. When this occurs, use tweezers to hold them in place.

Safety Precautions in the Lab

1. Work under a fume hood or in a well-ventillated area when processing gaseous or fuming materials.
2. Wear rubber gloves when weighing or preparing resin mixes. Some materials can cause dermatitis if allowed repeated contact with the skin.
3. Use acetone for cleanup operations. It is one of the best solvents from a health standpoint. Do not use toluene because it attacks rubber gloves.
4. Do not allow chlorinated solvents to come in contact with the skin. Repeated contact can cause a detrimental buildup in the body. These are materials such as carbon tetra chloride and trichloroethylene.
5. Never pour water into acid. When making chromic acid, pour the acid into the water, and pour slowly.
6. Cured urethane (rigid) foam can be very slippery on floors. If some should fall on the floor, gather it up immediately.
7. Metal items such as fixtures and clamps should never be put away while hot. Let them cool down first. If you have to leave the area, put a "hot" sign on them.
8. Keep hot catalyzed resins away from your face. They give off disagreeable vapors. Those containing epoxy curing agents such as metaphenylene diamine contain a brown dye that stains clothes unless held at arm's length. This particular group can be in the form of either tan to brown crystals or brown to dark brown liquids.
9. When you pressurize the chamber of a vacuum encapsulating machine, hang a precautionary sign on it. Also lock the vacuum

lever with a metal pin or other means to preclude the possibility of someone accidentally pushing it open.

10. Low-viscosity materials can permeate paper cups. When working with these materials, use poly beakers. If you must use a paper cup, wrap masking tape around it—or place the cup into a second cup. This will prevent contact with the hands.
11. When opening oven doors, keep your head back so as not to inhale any of the fumes that might be emitting at the time. Urethane foam, in particular, can be very noxious during the early stages of cure.
12. Keep all containers properly labeled.
13. Most labs have vinyl tile floors. When spillage occurs, wet with isopropyl alcohol. Do not clean with acetone or MEK. Vinyl tile is attacked by ketones and will be defaced if cleaned with acetone or MEK.

SPECIFICATIONS AND INSPECTION

Specifications

These are documents that describe materials and processes, among other things. For a bonding operation, the worksheet and the print will call out the process specification to be used for that particular operation. The technician then consults the specification to find out which adhesive formulation is to be used. The specification lists materials, proportions, cure cycles, and pot life. It also lists the material specification for each material, including all pertinent facts regarding the material including vendors and equivalent materials. Though several materials may be listed on the material specification, the callout on the process specification is specific.

Note: The author has taken the liberty of referring to specifications as *spex.* This is consistent with lab jargon. Also used in the book are other terms of lab jargon including "RTV" and "catalyzed." Actually, RTV means "room temperature vulcanizing" but is used in reference to silicone rubbers that cure at room temperature. "Catalyzed," in the technical sense, means that a catalyst, which only agitates or precipitates a chemical reaction but does not itself become part of the new compound, has been added to a liquid resin. In lab jargon, however, "catalyzed" is used to refer to any resin to which any type of curing agent has been added. Most curing agents used to solidify epoxies become part of the newly formed compound, but in lab jargon they are "catalyzed." Though some terms used around labs and shops are actually misnomers, they are well understood by all personnel and are, of course, expedient.

Inspection

Each operation on a worksheet must be bought off (approval-stamped) before the start of the next operation. It is the job of the inspector to make sure that each operation has been performed according to the pertinent spex and that the quality of the work is up to standards. This is a very important function that technicians appreciate. An oversight or minor error can be picked out quickly by the alert inspector, thus preventing the release of inferior work and embarrassment to the technician.

13 Thermoset Labs

THERMOSET LABS

Though much of the plastics work performed in an aerospace electronics lab deals with thermoset materials, a larger variety of thermoset processes and materials can be found in aircraft and aerospace plastics labs that deal with research and quality control in support of vehicle fabrication. Indeed, some of the large aircraft plants are engaged in both types of production. It is here in thermoset labs that a technician receives the best education in thermoset technology, especially in the area of reinforced plastics.

The technician in these labs prepares and tests specimens for tensile strength, compressive strength, flexural strength, impact strength, hardness, etc. In addition to tests for mechanical values, the technician also conducts physical tests on resins, pre-pregs, and cured laminates that include, among others, tests for resin content, resin solids, viscosity, volatiles, flow, and specific gravity. Mechanical specimens are prepared by vacuum-bag and laminating-press techniques. Laminate specimens are cut, routed, or otherwise machined to net dimensions.

Molding and casting specimens can be prepared in fixtures having net dimensions so that no further machining is required, Equipment used includes test machines, platen presses, balances (including analytical), ovens, muffle furnaces, calculators, viscometers, hydrometers, tensile routers, band saws, drills, durometers, hand tools, and impregnating apparatus. (Figure 13-1 shows miscellaneous lab equipment.)

Other duties of technicians in thermoset labs include adhesive bonding, preparation of surfaces for bonding (including preparation of etchant solutions), calculating for mechanical and physical values, and writing reports. Because of the broad scope of materials and processes in these labs, technicians are afforded excellent opportunities for advancement. A high percentage of technicians rise from hourly wages to salary in thermoset labs.

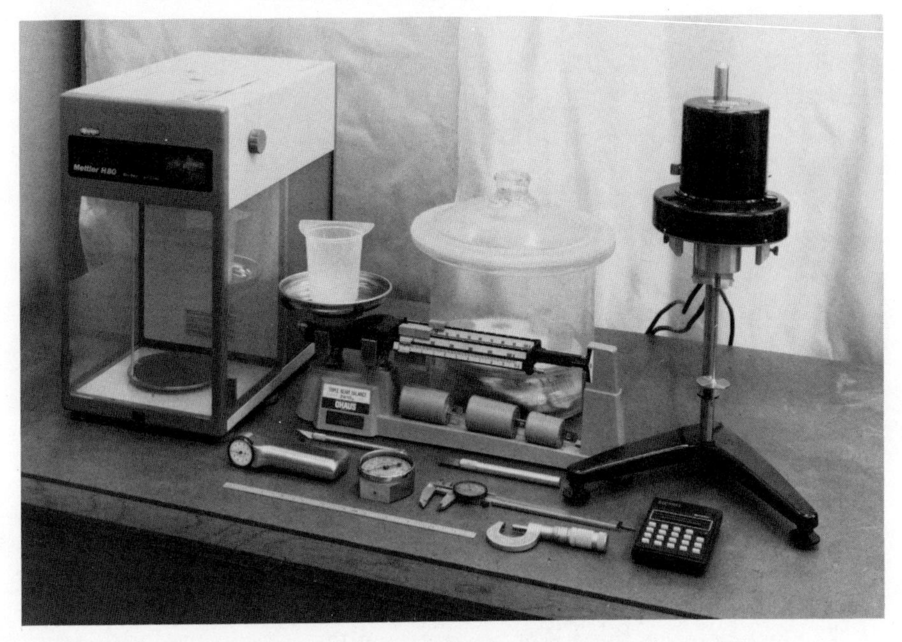

Figure 13-1. Miscellaneous lab equipment.

Laboratory Platen Presses

Reinforced plastics labs are normally equipped with several small presses that are used to pressurize and cure test laminates, lap shear specimens, and molding compounds (small molds). (See Figure 13-2.)

Test laminates can be either 4 in. × 4 in. stacks of pre-preg for flow tests, or they can be larger laminates prepared for research data. The plies are sandwiched between Teflon or Tedlar release film and flat aluminum plates called *cauls*. If the platens are not completely parallel to each other, a thin sheet of silicone rubber can be placed at the top and bottom of the assembly to take up the discrepancies.

Platen sizes usually range from 6 in. × 6 in. to 12 in. × 18 in. They are electrically heated and hydraulically operated. A water-cooled ram adapter can be fastened around the ram to prevent conduction of heat from the platens to the hydraulic oil. Heating the oil would result in a rise in pressure on the laminate.

Most lab presses are manually operated. A lever is pumped up and down to close the press and apply pressure, much like the action of a jack handle.

Drag weight, which is the dead weight of the lower platen assembly, is a factor when computing pressure settings. When the platens are brought to within ⅛–¼ in. of closing, the reading on the guage is the drag weight and is added to the calculated load. For low-pressure settings it is important to add the drag. For high pressure in the tons, drag is negligible and can be discounted.

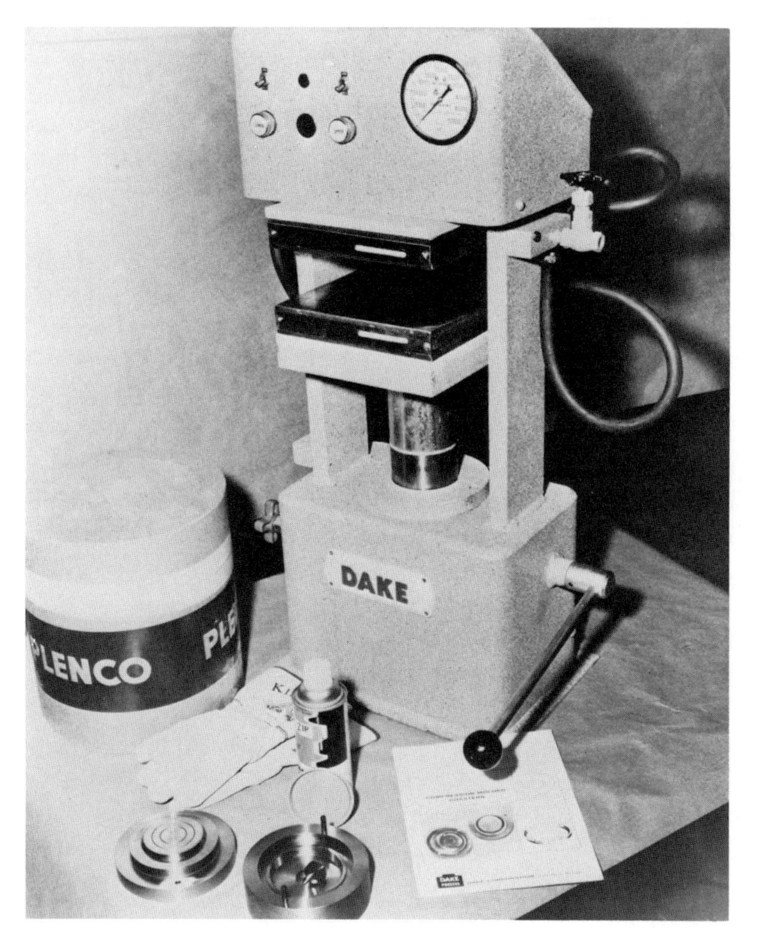

Figure 13-2. Small lab press and materials for a molding exercise. *(Courtesy of Dake Corporation)*

Operation of Presses

The following procedure applies to many manually operated lab presses currently in use.

1. Platens must be checked for alignment before each use by bringing them together. Any discrepancies noted should be reported before use of press.
2. Before turning on heat, turn on water and channel it through the ram adapter, making certain that no water is running through the platens. To do this, turn off platen water valve and also water inlet valve to other presses.
3. To turn heat on, turn plastic knobs behind platens in a counterclockwise manner. When red lights are on, platens are heating.
4. To turn heat off, screw in plastic knobs in a clockwise manner, until red lights go off.
5. To set platens for a specific temperature, turn knobs and let temperature rise to within 5°F of desired temperature. At this point, turn knobs in (counterclockwise) until the red lights go off. Temperature will creep up approximately 5° more.
6. At the conclusion of the cure, turn the switch off and if cooling under pressure is desired, open the platen valve, allowing water to run through the platens. Be sure pressure is maintained by the cranking lever enough to compensate for pressure loss due to contraction. When sample has cooled sufficiently, turn off water and turn on air to force out remaining water.

Gauge Setting for Pressure. Gauge settings for presses graduated in pounds or tons force are calculated as follows:

Example: Panel 10 in × 10 in. to be cured at 20 psi in press with drag of 10 lb.

$10 \text{ in.} \times 10 \text{ in.} \times 20^{+10} \text{ (drag)} = 2010 \text{ lb}$

For presses graduated in psi oil pressure the gauge setting (psig) is arrived at by multiplying the area of the laminate by the desired pressure (psi) then dividing by the area of the ram.

Example: Panel 10 in. × 10 in. to be cured at 15 psi on press having a ram diameter of 4 in. and a drag of 9 lb

$$\frac{10 \text{ in.} \times 10 \text{ in.} \times 15 \text{ psi}}{12.5 \text{ sq in}} = 120 + 9 \text{ (drag)} = 129 \text{ psig}$$

Use of Lab Scales. The small balances used in plastics labs must be kept free of liquids and powders to function properly. Spillage should be cleaned when it occurs. A build up of powders, in particular, can cause malfunction and erroneous weight readings.

Thick liquids are rather easy to transfer to the weighing container, but those with water-thin viscosities should be transferred along a vertically held spatula. Most powdered fillers can be placed in the weighing container without spilling by use of a wood tongue blade or spatula. Spillage of fillers is caused by trying to transfer too much at one time, or by rushing. There is no room for rushing in any type of lab where measurements and tests are precise. Efficient work speed is attained only by experience. Familiarity with the process at hand will allow a technician to work smoothly and efficiently without rushing. Powdered silica is the only filler that is difficult to transfer to the weighing container. This is because of its extremely low dry weight. When care is not used the powder seems almost to float off the transfer tool. For this type of filler, a small aluminum dish seems to be a better transfer tool than a spatula. For mixing, back and forth cutting strokes are best.

Tare Weight. This is the weight of the mix container and is used when weighing on balances having only one pan. The weight of the container is recorded before addition of the component materials. For balances having two pans, one container is placed on each pan. This maintains balance at zero, providing the containers are of equal weight. Some semiautomatic single pan balances can be used without recording tare weight. After the container has been placed on the weighing pan, a knob is turned to return the reading back to zero.

Weight Callouts. Care must be used when interpreting the various weight callouts in resin formulations. The simplest callout is PBW (parts by weight). It is normally used for formulations containing more than two ingredients. A number after each material listed indicates the exact weight (usually in grams) to be used in the mix. This is the basic formulation. To make larger batches, each weight callout is multiplied by the same number. If the batch is to be doubled, the multiplier is 2. If it is to be tripled, the multiplier is 3, and so on. Basic formulations do not always total 100, so that care must be used when relaying formulation information. Ten grams of curing agent in a 78-gram batch size is obviously not 10%; it is 10 PBW.

When the formulation contains just a resin and hardener, the hardener is usually called out as PHR (parts per hundred resin). A callout of 10 PHR means that 10 grams of hardener will be added to each 100 grams of resin. The 10 grams is actually 10% of the weight of resin, but not 10% of the mix. Straight percentage callout for material formulations is rarely used in plastics except for solutions made up of solvents and solid resins. A solution containing 10 grams of resin and 90 grams of acetone would be called a 10% solution.

Preparation of Lab Pre-Pregs

Fabrication of test specimens for developmental work often involves the preparation of pre-pregs. This consists of resin formulation and processing with lab-size equipment. Electrically operated machines such as the one shown in Figure 13-3, can be used for impregnation. Processing is similar to that used by manufacturers of production pre-pregs. Resin content is controlled by a combination of solvent content and doctor-blade adjustment. After the carrier is impregnated it passes between the two adjustable blades. Thc size of the opening determines the amount of resin solution deposited onto the carrier.

After the carrier has been impregnated, it is usually dried at room temperature for 30–60 min, then oven dried at 120–130°F. This removes most of the solvent from the pre-preg. Heat cycles actually vary depending on (a) resin system, (b) whether or not the pre-pregs will be B-staged, and (c) degree of B-stage. (See Figure 13-4 for picture of a lab oven.)

Lab pre-pregs can be used singularly as adhesive bonding films or in multiple plies for laminates. Cured laminates are machined into various physical and mechanical specimens.

Some lab pre-pregs, especially those used for one- to three-ply facings for honeycomb, are made by depositing a precise amount of resin on the carrier. For this system, the resin solution is weighed directly onto the carrier. The carrier is placed on a separator film such as Tedlar then

Figure 13-3. Lab-sized resin impregnating machine.

transferred to the balance. A flat piece of cardboard or other rigid material can be used on the weighing pan to provide a large enough area of support for the film and carrier. The film and support become the tare weight. After the specified amount of resin solution is weighed out, the solution is quickly spread out evenly over the carrier with a Teflon squeegee. The separator film is left on the layup during cure. This assures 100% transfer of the resin to the carrier.

Solvents are used in pre-preg formulations to facilitate distribution of the resin. A system designed to impregnate 100 sq in. of Kevlar with 50% resin (by weight) would only require a little over 4 grams of resin.

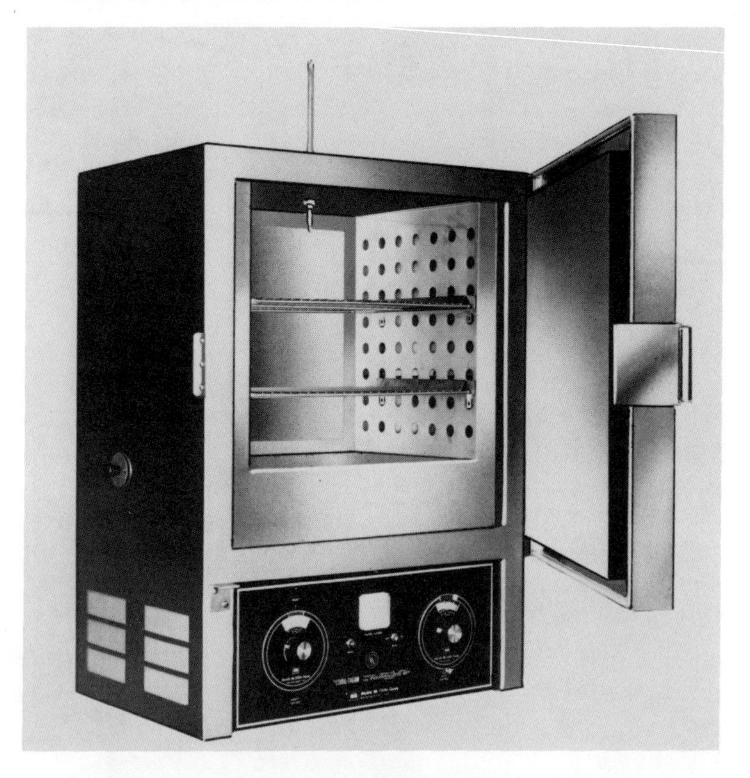

Figure 13-4. Electric oven. *(Courtesy of Blue M. Electric Co.)*

Obviously, this type of impregnation could not be made effectively without the use of solvents to extend the mass and lower resin vicosity.

Some processing observations are as follows:

1. Find out if there is a specific sequence to be followed when weighing and mixing the materials.
2. Powdered fillers should be dry and free of lumps.
3. Stir the ingredients slowly at first to avoid splashing. Paint mixers are excellent for mixing powdered resins and solvents. Metal containers are normally used with mix time 30–60 min.
4. Solvent-based mixes should be kept tightly covered when not in use to maintain the solid–solvent ratio. This is particularly true when the solvent is acetone because of its high degree of volatility. When mixing small batches of acetone–based solutions, the spatula should be weighed in with the component materials, so that when mixing is completed the solution can be weighed back

and the lost solvent replaced. Steel spatulas and metal cans or poly beakers should be used to prevent loss of solvent by absorption, which occurs when using wood spatulas and paper cups.

5. When pre-preg is being dried under a fume hood, it is important that no parting agents or sprays of any kind be allowed in the hood until the pre-preg has been removed.
6. Pre-pregs must not be dried in ovens used to process silicones. Most silicones will inhibit the cure or reduce the bond strength of other thermosets.
7. When pre-pregs come out of oven dry, they should be handled with clean white gloves for subsequent operations.
8. Glass cloth can be cut with sharp shears, but graphite cloth is easier to cut by using an aluminum template and a blade tool, such as a Stanley or Xacto knife. Graphite cloth has a tendency to unravel at the edges, and care must be used in handling and impregnating the patterns.
9. Some strong unidirectional fibrous materials such as Kevlar are difficult to cut. The task can be simplified by applying masking tape to one side of the cutting area, then using sharp shears to make the cut. The tape can be easily removed by wetting the adhesive side with MEK or acetone.
10. Racks for holding aluminum adherend panels and other plates can be made by notching a wood board as shown in Figure 13-5.

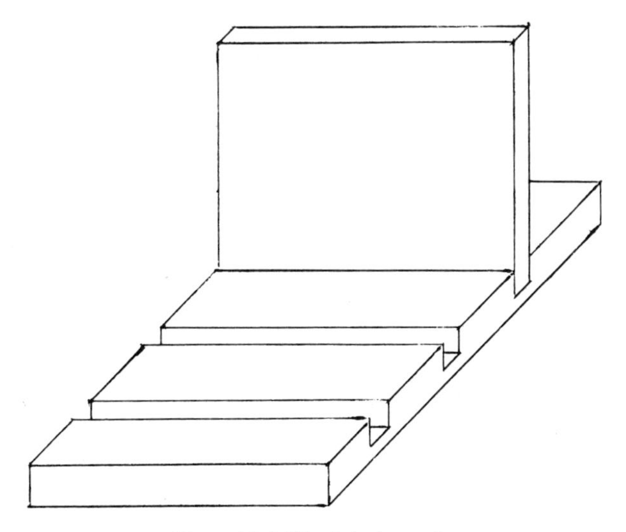

Figure 13-5. Wood drying rack.

The racks are especially useful for oven-drying adherends following the chromic-acid etching operation.

11. For precise weighings, add the materials slowly while gently tapping the container to determine the degree of resistance. When the weighing pan moves with a slight tap, it means that only slightly more material is needed to reach balance. At this point, additions can be made by drops to prevent an overrun.

Impregnation Exercise

Determine the weight of epoxy solution required to impregnate a strip of Kevlar fabric, 5 in. × 22 in., so that resin content will be 55%. The Kevlar weighs 4.4 grams and the solvent content of the resin solution is 20%.

Solution. Several methods can be used to find the weight of resin. Ratio and proportion is one of them. Just cross multiply, then divide. Note that the denominator is the × figure, and 45 represents % Kevlar (100-55). To find the weight of epoxy solution, divide the weight of resin by 0.80 (80%).

$$\frac{4.4}{45} = \frac{X}{55} \text{ or } \frac{45}{55} = \frac{4.4}{X}$$

$$= \frac{4.4 \times 55}{45X}$$

$$X = 5.37 \text{ grams of resin}$$

$$\frac{5.37}{0.80} = 6.71 \text{ grams of resin solution}$$

14 Physical Tests for Plastics

QUALITY CONTROL

Quality control means exactly what the words imply: controlling the quality of the product. There are many things that go into the overall control of quality, not the least of which is physical and mechanical testing of vendor materials to be sure that they are in compliance with order specifications. This chapter deals with some of the physical tests performed by plastics technicians, not only on incoming materials, but also on in-house production and research specimens, all of which help to control and optimize the end product. Most tests are for pre-pregs laminates made with pre-pregs, adhesive films, and resins.

Preparation of Pre-Pregs

Pre-pregs are fabrics and fibers that have been impregnated with a catalyzed resin system. Carriers such as fiberglass, high silica, graphite, carbon, and polyamide are passed through a dip tank containing a solution of resin and volatile solvents. Various devices are used to control the amount of resin pickup onto the carrier. The wet material then passes through a heated chamber where most of the solvent is flashed off. Further heating advances the resin to a predetermined state of B-stage. Though almost all pre-pregs are B-stage, there are some that remain in the A-stage even when exposed to moderate heat. They are tack free at room temperature and easy to work with. Shop life is extensive and cured properties predictable. An example of an A-stage pre-preg would be one with an allylic-dicumyl peroxide resin system. This type of system does not normally advance at temperatures below 200° F.

Tack and *drape* are also determined during the pre-preg operation. Tack is the degree of stickiness, and drape is the degree of droop and bend. The high-pressure laminator of flat laminates does not require any tack or drape in his or her raw materials, but the vacuum-bag molder needs both to allow the pre-preg to conform and adhere to the vertical surfaces and sharp contours of the mold.

When tack or drape is insufficient, the pre-preg can be heated with a heat gun to improve both conditions. Because too much heat will advance the resin and create problems, care must be exercised when using this technique. In some cases, a small amount of an appropriate solvent on the surface of the pre-preg can improve tack, but heating is more acceptable.

Physical and mechanical values of pre-pregs are more predictable than those obtained by wet-layup techniques. Pre-pregs are also much easier to work with than the sometimes messy wet-layup methods. As previously noted, some wet layups do have the advantage of curing with no applied heat or pressure. Because of this, the mold, which does not have to be very sturdy, can be built much cheaper.

Test Methods

Test methods used to control the quality of plastic materials are drawn up by the American Society for Testing and Materials (ASTM), an organization of professional people who publish standard methods for testing and classifying materials used by the industry. Products made for the United States Air Force and Navy are controlled by tests listed under Federal Publication LP406-B.

Testing for Flow (Pre-Pregs)

1. Cut enough pieces of material 4 in. × 4 in. to make approximately a 20-gram specimen. All samples for testing should be cut on the bias. Stack evenly, weigh, and record.
2. Place between release film such as Teflon cloth, and sandwich between caul plates.
3. Place in a press and apply the specified heat and pressure. Full cure is not needed. Specimen is ready when it is tack-free, and there is no more movement of resin.
4. Upon completion of the cycle, remove laminate assembly from the press, and allow it to cool to room temperature.
5. Break off and/or scrape off all flash, being careful not to remove any of the cloth strands.
6. Weigh back and calculate.

$$\frac{\text{original wt} - \text{final wt}}{\text{original wt}} \times 100 = \%\ \text{flow}$$

Testing for Volatile Content

This is a test to determine the amount of volatile material in a pre-preg.

1. Cut out specified number of specimens 4 in. × 4 in.
2. Puncture a small hole in one corner of each specimen.
3. Weigh the specimens on an analytical balance to the nearest thousand.
4. Using a paper clip or soft wire hook, suspend the specimen from a rack in a circulating oven.
5. Cure per specification.
6. Cool in a desiccator for 10 min and weigh back and calculate.

$$\frac{\text{original wt} - \text{dry wt}}{\text{original wt}} \times 100 = \%\ \text{volatiles}$$

Testing for Resin Solids (Glass Carriers)

Resin solid content is the amount of resin remaining in a pre-preg after all volatile matter has been removed.

1. Use the same specimens that were oven-cured for the volatile test.
2. Transfer specimens to a muffle furnace heated to 1050°F for 30 min, or until all traces of resin have disappeared. Excessive temperatures with fiberglass specimens could melt the glas, 1550°F is used for high silica and asbestos.
3. Place crucibles in a desiccator until cooled to room temperature.
4. Weigh back and calculate.

$$\frac{\text{wt of devolatilized sample} - \text{wt after burn-off} \times 100}{\text{wt of devolatilized sample}} = \%\ \text{resin solids}$$

Resin Pickup Test

This is a test to determine the amount of resin in a pre-preg, including volatiles.

1. Weigh specimens of the pre-preg, as received.
2. Place in a tared crucible and transfer to a muffle furnace until all but the fabric is burned off.

3. Remove from the muffle and place in a desiccator until specimens reach room temperature.
4. Weigh back and calculate.

$$\frac{\text{original wt} - \text{final wt} \times 100}{\text{original wt}} = \%\ \text{resin pickup}$$

To compute resin pickup of pre-pregs having combustible carriers, immerse 2 in. × 2 in. specimens in a suitable solvent until all of the resin has been dissolved. Dry thoroughly and weigh back. Use the above formula for calculation.

Resin Content of Cured Laminates (by Burnout)

When it is desirable to know the resin content of a cured laminate, specimens can be taken from trim areas. When the entire laminate is functional, a separate test panel can be processed along with the production part.

1. Select convenient sizes to fit into crucibles.
2. Record weight of specimens and burn off the resin in a muffle furnace per the method used for computing resin solids of pre-pregs.
3. When cooled to room temperature, weigh back and calculate.

$$\frac{\text{original wt} - \text{final wt}}{\text{original wt}} \times 100 = \%\ \text{solids}$$

For asbestos laminates, the following formulation must be used because of the water of crystalization.

$$\frac{\text{original wt} - \text{final wt}}{\text{original wt}} \times 1.16 \times 100 = \%\ \text{solids}$$

Chemical Testing for Resin Content

This is the method often used in the determination of resin content of graphite-epoxy and carbon-epoxy pre-pregs and cured laminates. For pre-pregs, the specimen is devolatilized before testing. Testing is conducted as follows:

1. Weigh the specimen then place in a 250-ml beaker. Specimen size should be approximately 2 in. × 2 in. for pre-pregs and ½ in. × 1 in. for laminates.
2. Pour 100 ml of concentrated nitric acid (HNO_3) into the cold beaker.
3. Place the beaker on a hot plate, and stabilize the temperature of the acid at 150–160° F.
4. Digest the specimens for 1½–2 hours or until all the fibers have been separated as determined by visual examination.
5. Filter in a Buchner funnel and vacuum flask using a tared glass-fiber filter. A crucible can be used if preferred, but the wider filter provides a larger work area.
6. Rinse thoroughly, first with distilled water, then acetone.
7. Dry in an oven, then transfer to a desiccator for 10 min. Weigh back and calculate.

$$\% \text{ resin} = \frac{\text{original wt} - \text{wt of fibers}}{\text{original wt}} \times 100$$

Note: The rinse water containing nitric acid should be poured into a properly labeled glass bottle, then neutralized with a base such as ammonium hydroxide when the test has been concluded. The rinse acetone, which should contain no more than a slight trace of acid, can be disposed of in a spent solvent can. Under no circumstances should the acetone rinse be poured into the water rinse bottle. Any concentration of nitric acid in the presence of acetone can cause a vigorous reaction. This operation should never be performed without some type of face or head protection.

Resin Content by Measurement

The following procedure can be used to determine the resin content of epoxy laminates having carriers that cannot be subjected either to burn-off or acid digestion.

$$W_1 \text{ (grams/sq in.)} = \frac{\text{wt of laminate (grams)}}{\text{area of laminate (sq in.)} \times \text{number of plies}}$$

$$W_2 = \text{grams/sq in. of unimpregnated fabric}$$

$$W_3 = W_1 - W_2 = \text{grams/sq in. of resin}$$

$$\% \text{ resin content} = \frac{W_3}{W_1} \times 100$$

To obtain the weight of the unimpregnated fabric, use MEK to extract the resin from a 4 in. × 4 in. piece of pre-preg.

Specific Gravity

There are several methods to determine specific gravity of cured laminates. Probably the most common is the water displacement method, as follows:

1. Dry the specimen in an oven. Cool to room temperature in a desiccator, then weigh on an analytical balance.
2. Mount a platform over the weighing pan in such a manner that there is no contact with the pan. Fill a 250 ml beaker with distilled water to within ½ in. of the top, and place beaker on the platform.
3. Wrap a tared fine-wire filament around the specimen, immerse it in the water, and move it up and down until all air bubbles have been removed from the specimen. Make a loop on the other end of the wire and hang it on the pan frame hook. Check to be sure that the specimen is below the surface of the water and free from contact with the beaker. Weigh and record. If balance has two pans, suspend on equal-length wire from each side. This will eliminate tare weight.

$$\text{sp gr} = \frac{\text{wt in air}}{\text{wt in air} - \text{wt in water}}$$

If specimens are porous, they should be degassed before weighing in water. This is accomplished by evacuating in a vacuum flask containing distilled water: the specimen is ready when all air bubbles have disappeared. It is also a good practice to boil the water that the specimen is to be weighed in, in order to remove the air from the water.

Trichloroethylene is another liquid sometimes used to determine specific gravity by displacement. For this method, calculate as follows:

$$\frac{\text{wt in air} \times \text{sp gr of trichlor.}}{\text{wt in air} - \text{wt in trichlor.}}$$

Specific gravity of liquids is measured by hydrometers.

Viscosity

Viscosity is a measure of the degree of flow of a liquid. Thick liquids such as molasses have a high viscosity; thin liquids such as water have a low viscosity. If a resin has started to advance, a test will show a higher viscosity than normal.

Various types of apparatuses are used to measure viscosity. Zahn cups are frequently used to test low-viscosity materials. After the cup is filled, the liquid escapes through an opening in the base of the cup. The time required to empty the cup is recorded and used as a comparison figure. Tests can also be measured in drops per min. Zahn cups are available with different-size openings.

One of the most widely used instruments for measuring the viscosity of resins, the Brookfield Viscometer measures the resistance of the resin to a rotating metal spindle. Readings are in poises and centipoises.

Fiber Volume and Void Content

Several methods are used to determine fiber volume and void content in graphite-epoxy laminates. The following methods are among those current in use.

$$\%\ \text{fiber volume} = \frac{\text{fiber wt}}{\text{laminate wt}} \times \frac{\text{laminate sp gr}}{\text{fibers sp gr}} \times 100$$

$$\%\ \text{void content} = \frac{\text{laminate wt}}{\text{laminate sp gr}} - \left[\frac{\text{resin wt}}{\text{resin sp gr}} + \frac{\text{fiber wt}}{\text{fiber sp gr}}\right] \Big/ \frac{\text{laminate wt}}{\text{laminate sp gr}} \times 100$$

Specific gravity of the resin and fibers is furnished by vendors of the pre-pregs. Resin weight and fiber weight are computed after acid digestion of the resin.

15 Mechanical Tests for Plastics

TEST MACHINES

Modern test machines (see Figure 15-1) can be used for a variety of mechanical tests. Standard machines have a movable member that either can pull a specimen to destruction, as in tensile and lap shear tests, or can push a specimen to failure, as for compressive and flexural tests. The amount of stretching that a specimen undergoes during a test can be automatically plotted on a chart located on the machine.

Grips for specimens pulled in tension are self-aligning. (Grip surfaces are slightly serrated to prevent slippage.) Specimens are mounted securely in the jaws of the fixture, and the force is applied at a constant rate and at a predetermined speed.

An electrical device called an *extensometer* is attached to tensile specimens when an accurate stress–strain curve is required. The extensometer is removed from the specimen before the specimen is pulled to destruction.

Depending on the requirements, specimens are tested at room temperature, elevated temperatures, and low temperatures. An environmental chamber is mounted around the test fixture when high or low temperatures are used. Lap shear specimens are sometimes pulled while in a liquid nitrogen (LN_2) bath, which has a temperature of approximately $-320°F$.

The moving member of the test machine travels at a rate of 0.020 in./min to 20 in./min. Many tests are conducted at a crosshead speed of 0.05 in./min. Tests that are conducted at a much higher speed than normal will show erroneously higher readings.

Tensile Test (ASTM 638)

Tensile strength is a measure of the resistance of a material to stresses pulling in opposite directions. A taut guitar string is an example of ten-

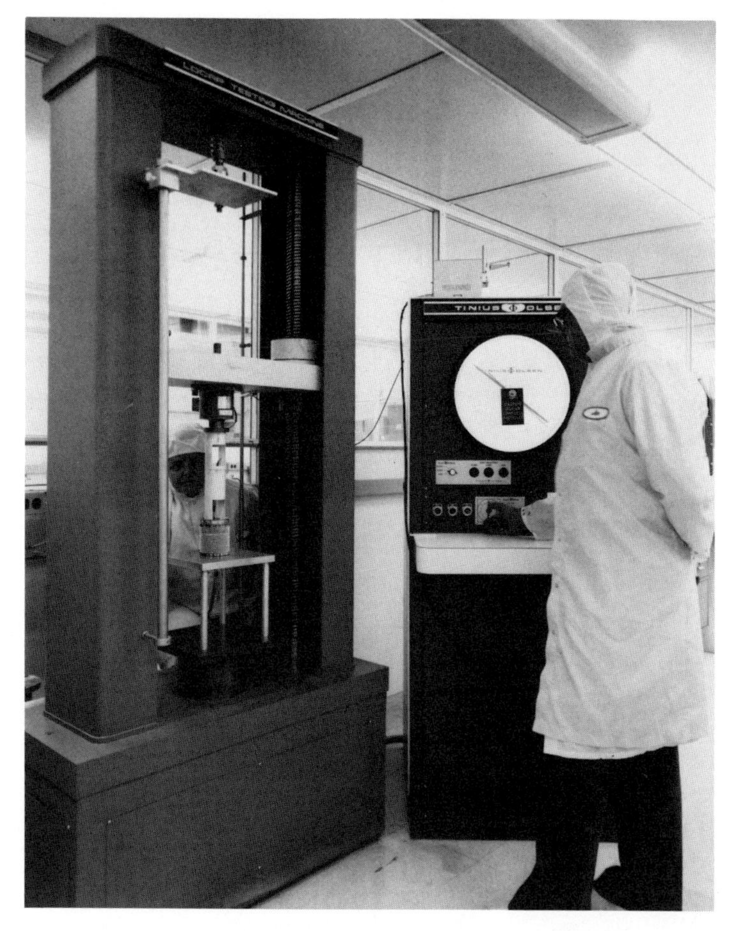

Figure 15-1. Testing machine with compressive test in progress. *(Courtesy of Hughes Aircraft)*

sile stress. When tensile specimens are tested they are said to be *pulled in tension.*

Fiberglass-reinforced specimens are usually machined to the shape of a dumbell, as noted in Figure 15-2. Specimens containing high-modulus boron or graphite are usually machined with parallel sides (also noted in Figure 15-2). The tapered end tabs bonded to the straight-sided specimens prevent indentation of the specimen by the serrated grip surfaces. Indentation with this type of specimen could cause premature failure.

Slippage of the specimen in the jaws is sometimes encountered at the

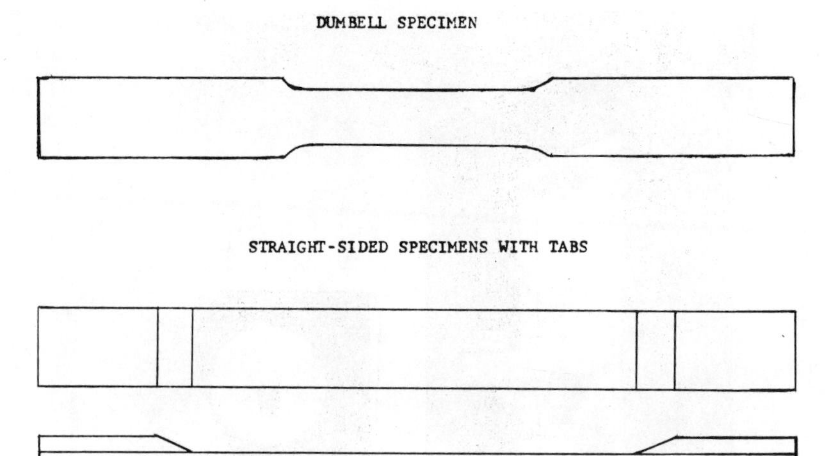

Figure 15-2. Tensile specimens.

start of a tensile test, particularly when the gripping surfaces are not serrated. A piece of fine grit sandpaper can be used to prevent this occurrence. In stubborn cases, two pieces of sandpaper can be bonded together with double-back tape so that there is grit facing both the specimen and the jaw surface.

Molds having the dumbell shape are used in the preparation of specimens for evaluation of molding compounds and casting or potting materials. Fiberglass-reinforced laminates are routed to the dumbell shape on a tensile router. Rubbers can be punched out with a steel rule die.

$$\text{Tensile strength (psi)} = \frac{\text{load (lb)}}{\text{cross-sectional area (sq in.)}}$$

Load is the ultimate load to cause failure of the specimen. *Cross-sectional area* is the width × the thickness of the neck area. If there is a variation in dimesions, measurement is taken at the smallest area.

Tensile Shear Strength

This test, more commonly known as the lap shear test, is used to measure the strength of adhesive-bonded joints when stressed in shear. The adhesive can be in the form of a paste, dry, unsupported tape, or B-staged fabric.

Specimens are prepared by bonding one adherend to another in a lap joint. The adhesive is positioned on the treated surface of one of the adherends then the other adherend is placed over it to the desired lap length and in a coinciding direction. Aluminum adherends are widely used for evaluation of supported B-stage adhesives. They are available as panels consisting of four individual adherends lightly held together. Two panels are assembled in a bonding fixture with a strip of adhesive sandwiched between the adherends in the bond area. The assembly is then placed in a platen press where the adhesive is cured under heat and pressure. When ready for testing, the adherends are separated by hand or on a band saw. Excessive adhesive squeezed out around the side edges of the specimens should be removed before testing because it can cause the specimens to fail at erroneously higher loads. When failed specimens have adhesive remaining on the entire bonding area of both adherends, the failure is said to be 100% cohesive, the optimum condition. The poorest condition occurs when all the adhesive remains on one adherend while the other is as clean as a whistle. This is called 100% adhesive failure and could also be called 0% cohesive failure. Of similar consequence would be a failure in which half of each bonding surface is still covered but not in a mating alignment. In other words, if the failed specimen were fitted back together, there would be no adhesive-to-adhesive contact, only adhesive-to-bare-metal. Failure of this type usually falls somewhere between these two extreme conditions. (See Figure 15-3.)

Lap shear specimens are pulled in tension the same as in tensile test. The fixture must be self-aligning to avoid stress in peel. Some circumstances actually require mechanical support of the adherends in the bond areas to prevent failure in peel.

$$\text{Lap shear strength (psi)} = \frac{\text{load to fail specimen (lb)}}{\text{bond area (sq in.)}}$$

Modulus of Elasticity (E)

This is the measurement of the stiffness of a rigid material when it is subjected to tensile, flexural, compressive, or torsional forces. When a specimen is pulled in tension (tensile), it gets longer and narrower. If

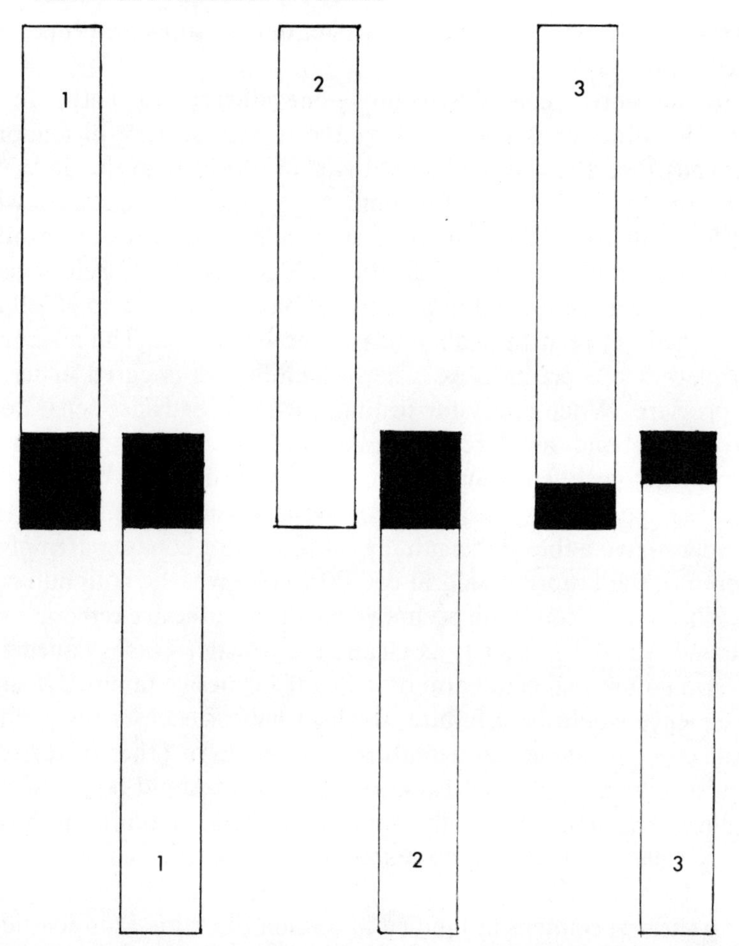

Figure 15-3. Adhesive failures. (1) 100% cohesive, (2) 100% adhesive or 0° cohesive, (3) also 100% adhesive.

the load is removed, it will recover its original dimensions, providing it has not been stressed beyond its *elastic limit,* the load beyond which the specimen starts to undergo permanent distortion. Permanent distortion is known as *permanent set,* or *permanent deformation.* Figure 15-4 shows a typical stress–strain curve plotted during the testing of a specimen. The straight-line portion represents all loads within the elastic limit. Modulus can be computed anywhere along this line. All loads on the curved portion beyond the yield load will cause permanent deformation. Hooke's Law states that when a load is applied to a material, the stress is directly proportional to the strain–within the elastic limit. If the modulus is computed at three different points along the straight-line portion of the curve, the readings should all be the same.

$$\text{Modulus} = \frac{\text{stress (psi)}}{\text{strain (in./in.)}}$$

$$\text{Stress (psi)} = \frac{\text{load (lbs)}}{\text{cross sectional area (sq in.)}}$$

Strain is read directly from the chart.

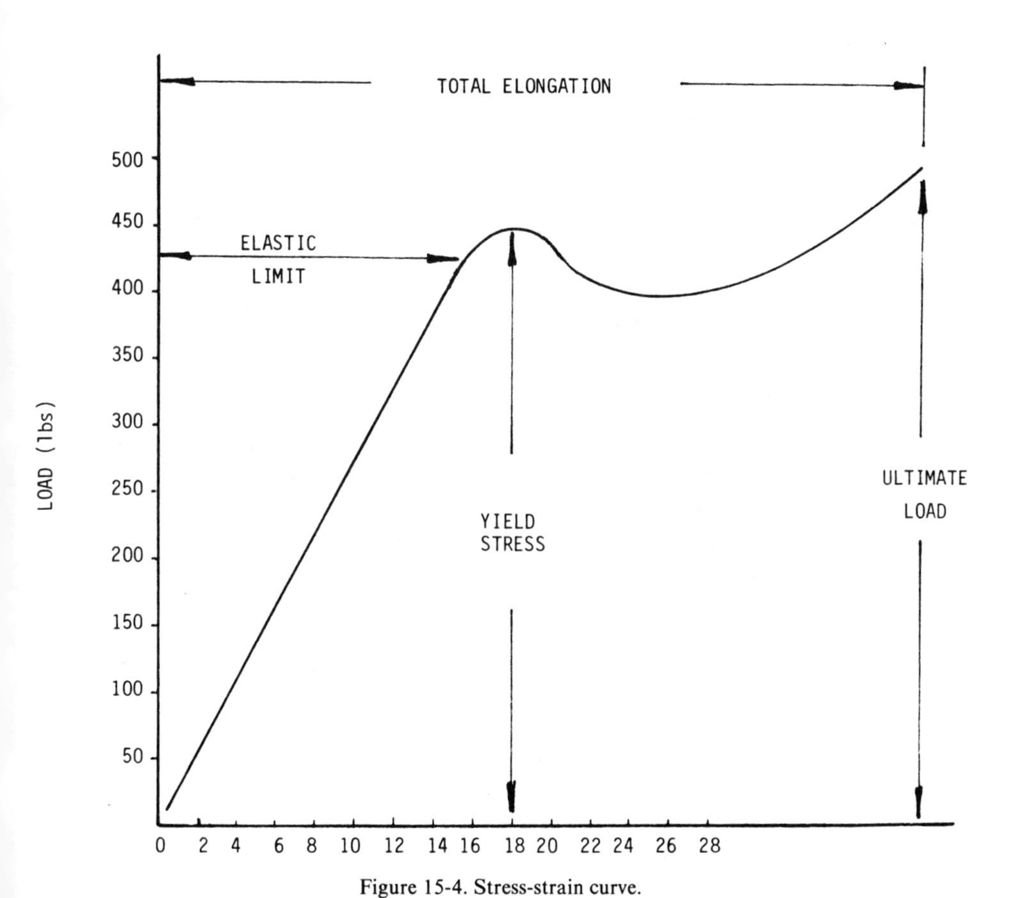

Figure 15-4. Stress-strain curve.

Example: Compute the tensile modulus with the load at 225 lb. The specimen is 0.500 in. wide and 0.125 in. thick in the neck area.

Answer: Cross-sectional area = 0.5 × 0.125 (0.0625 sq in.) Strain at 225 lb is 0.008 mil.

$$\frac{225}{0.0625} = 3600 \text{ psi}$$

$$\text{Modulus} = \frac{3600}{0.008}$$

$$\text{Tensile modulus (E)} = (450{,}000 \text{ psi})$$

Elongation

Elongation is the distance that a specimen stretches when pulled in tension. For rigid materials, the stretch is short. For rubbers, it can be many times the length of the specimen. Materials that can be stretched at least twice their normal length are known as *elastomers.*

$$\text{Total elongation (\%)} = \frac{\text{ultimate length} - \text{original length} \times 100}{\text{original length}}$$

Flexural Strength (ASTM D790)

Flexural stress is a combination of tensile and compressive forces. When you sit on a park bench, you are exerting flexural stress on the bench.

A specimen to be tested is placed on a fixture having a supporting upright near each end. The load is then applied directly downward and in the center of the specimen.

$$\text{Ultimate flexural strength (psi)} = \frac{3pl}{2bd^2}$$

where

p = load to cause failure (lb)
l = span from one upright to the other
b = width
d = thickness

Compressive Strength (ASTM D695)

This test is a measurement of the ability of a material to withstand compressive forces.

Specimen size is ½ in. × ½ in. × 1 in. for blocks and ½-in. diameter × 1 in. for cylinders. Since plastic products seldom fail from compressive loading, the compressive strength of plastics is of comparatively little design value. The data, however, are useful in the determination of the overall strength of different materials.

$$\text{Ultimate compressive strength (psi)} = \frac{\text{load to cause failure (lb)}}{\text{cross-sectional area (sq in.)}}$$

Izod Impact (ASTM D256)

IZOD impact test determines the ability of a material to resist a sharp blow. Specimen size is usually ⅛ in. × ½ in. × 2 ½ in. A notch is cut on the narrow face of the specimen, which is conditioned according to procedure A, ASTM D618.

The specimen is clamped at the base of the machine with the notch facing the direction of impact. The pendulum is released, and the reading is indicated by a pointer that is pushed along by the pendulum in its follow through.

The IZOD test is not necessarily a reliable indicator of overall toughness and impact. Materials such as nylon and acetal are among the toughest plastics but register low on the notched IZOD impact test because they are notch sensitive.

Indentation Hardness (Barcols)

A Barcol is a hand instrument used to measure the hardness of thermoset laminates. Pressure on the instrument causes a conical needle to penetrate the surface of the part or specimen. Readings taken in various areas can also indicate the degree of cure and density. A combination of high and low readings indicates nonuniformity with inclusion of resin-rich pockets. Readings range from 0 to 100.

INDENTATION HARDNESS (Durometers) (ASTM 2240)

Durometers are also small hand instruments used to measure indentation hardness of rubbers and plastics in somewhat the same manner as the Barcol tester.

Test specimens should have a minimum thickness of at least 0.25 in., though thinner specimens can be stacked for a reading. Materials harder than 50 Shore D should have a minimum thickness of 0.12 in.

Durometer readings should not be taken on an uneven, irregular, or coarsely grained surface. Round or cylindrical surfaces can be tested by "rocking" the Durometer on the convex surface and observing the maximum reading, which occurs when the indentor is aligned with the axis of the roller.

The temperature at which the test is made may have a significant effect on the reading, depending on the material involved. All readings taken at other than normal ambient temperatures should be recorded with a notation stating the actual temperature of the test.

Pressure of application of the Durometer should be sufficient to insure firm contact between the flat bottom of the Durometer and the test specimen.

All Durometer scales are graduated 0–100 with all indentors protruding 0.10 in. Readings are taken within 1 sec after firm contact between the flat bottom and the specimen has been established. The dial hand may gradually recede on materials such as nitrile rubber that exhibit cold flow or creep.

REFERENCES

"Standard Tests of Plastics." *Bulletin GIC,* 8th Ed. Celanese Plastics Company, Chatham, N.J., 1977.

Driver, W. E., *Plastics Chemistry and Technology.* Van Nostrand Reinhold Co., New York, 1979.

Shore Technical Bulletin. Shore Instruments and Manufacturing Company, Jamaica, New York, 1979.

16 Reference Tables and Guides

MATERIALS GUIDE

Cellulose Acetate

1. Good stiffness and impact strength. High gloss, good clarity, and color range.
2. Lowest priced cellulosic, but somewhat more water absorbent than others. Also has higher specific gravity.
3. Not suitable for outdoor use.
4. Available in sheets, rods, tubes, molding compounds.
5. Methods of processing include injection molding, extrusion, and some compression molding. Has excellent formability for thermoforming.

Ethyl Cellulose

1. Outstanding toughness, even at subzero temperatures.
2. High tensile strength, good rigidity, good moldability.
3. Processed by injection molding, extrusion, and thermoforming.
4. Coloring is good but confined to transparent, translucent and opaque shades.

Cellulose Acetate–Butyrate (Figure 16-1)

1. Good weathering characteristics. High resistance to water. Used for signs, advertising displays and other outdoor applications.
2. Good impact strength at low temperatures; good moldability and dimensional stability.
3. Used in injection molding, extrusion, and thermoforming. Good clarity, color range, and surface luster.
4. Has a mild rancid-butter odor that is somewhat noticeable indoors.
5. Available in sheets, tubes, molding compounds.

Figure 16-1. Clear dome made of cellulose acetate butyrate. *(Courtesy of Eastman Chemical Products, Inc.)*

Cellulose Nitrate

1. Excellent toughness and dimensional stability. Low water absorbtion.
2. Very flammable. Poor resistance to heat and sunlight.
3. Not used in processing requiring heated molds or other heated equipment.
4. Used mostly as base for lacquers and cements.

Cellulose Propionate

1. Good weathering properties and shock resistance.
2. Properties are similar to cellulose acetate butyrate. Both materials are tougher than cellulose acetate and somewhat easier to process.

Polyethylene

1. Member of a group called polyolefins.
2. Manufactured in high-density and low-density forms, depending on end use.
3. Leads all other plastic materials in volume produced.

4. Has good low-temperature toughness, low water absorbtion, excellent chemical resistance, and good flexibility at subzero temperatures.
5. Available in sheets, rods, tubes, and molding compounds.
6. Processing includes injection molding, extrusion vacuum forming, blow molding, and coating.
7. Products include squeeze bottles, toys, milk bottles, films for packaging, food and chemical containers.
8. Is lightweight with a natural white, waxy appearance. Thin films, however, are clear, transparent, and glossy.

Polypropylene

1. Another major member of the polyolefin family.
2. One of lightest plastic materials: floats in water.
3. Is tough, durable, water resistant. Has excellent electrical properties, even at elevated temperature. Has great flex life.
4. Has excellent hinge properties. Hinge is molded in as part of the container or binder.
5. Exceeds polyethylene in heat resistance and rigidity.

Polystyrene (Styrene)

1. Has outstanding transparency, hard surface, excellent electrical properties, low moisture absorbtion, and outstanding colorability.
2. Is brittle and does not have good resistance to weathering. Brittleness can be overcome by addition of butadiene, a synthetic rubber. This addition produces high-impact styrene with marked increase in physical properties.
3. General-purpose varieties have a metallic sound when dropped or tapped.
4. Major products include packaging containers, housewares, and toys. Styrene foams are finding increasing demand as insulation and flotation materials.

ABS

1. Material is formed through simultaneous polymerization (curing and hardening) of three materials: acrylontrile, butadiene, and styrene.

2. It is a tough, rugged plastic with good heat resistance.
3. Not transparent, but has good colorability. Has good toughness at low temperatures as well as good scratch and wear resistance.
5. Products include football helmets, boat hulls, luggage, radio and television housings.

Polyvinyl Chloride (PVC)

1. Water and chemical resistant, good weather resistance and strength.
2. Will not support combustion.
3. Properties are altered by the addition of modifiers to make end product tough and rigid or soft and flexible. Flexibility is developed by addition of plasticizers (softening agents).
4. Processed mostly by injection molding, extrusion, blow molding, calendering, and coating.
5. Products include upholstery materials, and waterproof fabrics such as shower curtains; also pipe fittings, chemical storage tanks, floor covering, packaging films, garden hose, and wire insulation.

Polyvinyl Acetate

1. Used primarily as emulsions for adhesives and paints. Also used as a chewing-gum base and as coatings for textiles.
2. Not suited for processing by molding because of inherent softness.

Polyvinyl Chloride-Acetate

1. Water and chemical resistant, but more soluble in organic solvents than PVC.
2. Joined with asbestos to form large market in floor tile and linoleum. Fabrication is by calendering.
3. Has been used extensively in the manufacture of phonograph records.

Polyvinylidene Chloride

1. Soft and flexible, transparent, odorless, tasteless and tough.
2. Low water and vapor transmission.

3. Major use is in films for food wrapping.
4. Products are prepared by extrusion, molding, and calendering.

Polyvinyl Butyral

1. Best known as the adhesive interlayer for safety glass, such as that used in automobile windshields.
2. Has outstanding clarity and good impact strength.

Polyvinyl Alcohol (PVA)

1. Well known to the plastics laminator as PVA release and bagging film. PVA, being water soluble, can also be sprayed or brushed on molds as either a direct release film or to form a contour-fitting vacuum bag. With the latter, thickness of the laminate must be considered, so that when the bag is applied it will have the proper fit.

Polyvinyl Formal

1. Used primarily in the manufacture of coatings for wires.
2. Can be formulated to have good toughness, flexibility, chemical resistance, and resistance to heat when modified with other resins.

Polytetrafluorethylene (TFE)

1. Member of fluorocarbon family, which is the most chemically inert of all plastics (not affected by chemicals or solvents).
2. Remains flexible at −400°F and remains stable to 500°F. Service temperature is to 390°F.
3. Has lowest coefficient of friction of any known solid, accounting for its no-stick properties.
4. Looks and feels like wax.
5. Color is opaque white, but thin sections are translucent.
6. High melt point makes processing difficult. At room temperature, however, TFE powders can be compacted under high pressure to a solid mass and then, heated to 700°F, fused to a solid form.
7. Products include gaskets, seals, O-rings, electrical insulation and nonstick coatings in cookware and processing equipment.
8. Has a high specific gravity for a plastics material.

Fluorinated Ethylene-Propylene (FEP)

1. Very much like TFE except that FEP has a lower melt point and melt viscosity, allowing it to be extruded and injection molded.
2. Has excellent heat aging properties.

Polychlorotrifluoro Ethylene (CTFE)

1. Another member of the fluorocarbon family.
2. Unlike TFE, it can be processed by extrusion, injection molding, and compression molding.
3. Like TFE, it has chemical inertness, high temperature resistance, and low temperature flexibility. Some swelling is encountered with some organic solvents, but the effects are of little consequence.
4. Crystal clear in thin films; good resistance to weather, colorability, impact strength, and abrasion resistance.
5. Used in printed circuits, chemical-processing-equipment closures for chemicals, valves, fittings, and pipe seals.
6. Service temperature is −400°F to 390°F.
7. Like all fluorocarbons CTFE is an expensive material.

Polycarbonates

1. Superior plastic for engineering. Outstanding toughness, long life under extreme conditions, excellent machining properties.
2. Serviceable from −220°F to +250°F.
3. Good transparency, good electrical properties, superior dimensional stability, self-extinguishing.
4. Processed by extrusion, injection molding, and blow molding. Also furnished in sheets, rods, and tubes.

Polyamide (Nylon)

1. Tough, strong, stable, abrasion resistant, heat resistant (to 250°F) and has low coefficient of friction.
2. Resistant to most organic solvents and most chemicals.
3. Used extensively for small industrial parts because of thin-wall strength. Other uses include ladies wear, tents, carpets, drawer slides, gears, housings, combs, fishing lines, and textiles.
4. Supplied in all standard forms and processed by most standard equipment.

Acetals

1. Compares with polycarbonates and nylons in structural applications.
2. Strong, tough, stable, rigid, good moldability.
3. Not attacked by hydrocarbons. Resists most chemicals. Service temperature to 220° F.
4. Used in sprinklers and pump assemblies because of extremely low moisture absorbtion. Other uses include zippers, valves, and housings.
5. Methods of fabrication include injection molding, extrusion, and thermoforming.

Polyurethanes

1. Produced as flexible, rigid, and semirigid foaming materials. Flexible foams are made in slabs and find wide use as foam rubber in furniture, mattresses, cushions, and safety applications. Rigid and semirigid urethane foams are finding expanded use in structural applications where the potential is unlimited. Urethane rubbers and foams are used by the aerospace industry to encapsulate electronic components to protect them against shock, vibration, and moisture.
2. Rigid elastomeric urethane rubbers are among the most abrasion resistant of all plastic materials. Because of this and their outstanding toughness, they are used as solid tires on industrial vehicles. They also find wide use as bumper guards and safety padding for automobiles.
3. Urethane coatings are tough, hard, flexible, and chemical resistant.
4. Urethanes have an inherent problem of flammability, but it is hoped that research will solve this problem soon, so that the full potential of this promising material can be realized.

Acrylics (Methyl Methacrylate)

1. Outstanding clarity and light transmission, outstanding colorability.
2. Excellent for outdoor applications; strong, rigid, low moisture absorbtion, good dimensional stability.
3. Attacked by some chemicals and organic solvents.

4. Can be molded and extruded, but largest business comes from production of sheets for outdoor signs. Thermoforming is another method of fabrication.
5. Applications also include shower doors, automobile instrument faces, light diffusers, and display fixtures.

Polyphenylene Oxide (PPO)

1. Has outstanding electrical properties, good thermal stability, high heat distortion point (375°F), good toughness, and low water absorbtion.
2. Requires molding temperature from 550 to 600°F.
3. Finds use in battery cases, electrical housings, and printed circuits.
4. High temperature resistance and low water absorbtion make PPO excellent for applications involving the use of hot water.

Polysulfone

1. Rigid and strong, wear and heat resistant (to 300°F).
2. Used as gasoline-pump and valve parts because of resistance to oil and gasoline. Also used as electrical insulating material in high-heat applications.
3. Processed by conventional thermoplastic processing equipment.

Ionomer

1. Has good combination of extreme toughness, solvent and chemical resistance, and low-temperature impact strength.
2. Application include golf balls, food packaging, soles for sport shoes, and blow-molded bottles.

Phenoxy

1. Crystal clear, tough, resistant to most chemicals, and has low heat expansion. Highly resistant to transmission of gases.
2. Soluble in some solvents. Can be solvent-cemented.
3. Applications include sports equipment, pipes for crude oil, and packaging.
4. Easily processed with standard equipment.

Polyallomer

1. Member of the polyolefin family.
2. Abrasion resistant, chemical resistant, good colorability, outstanding hinge properties, good strength and toughness.
3. Good use is made of its hinge properties in products such as luggage, casings, and book bindings.

Polyphenylene Sulfide

1. Has high melting point (550° F), outstanding chemical resistance, thermal stability and fire resistance of the polymer.
2. There are no known solvents below 375° F.
3. Has high stiffness and good retention of mechanical strength at elevated temperatures.
4. Can be used for slurry coatings, fluidized bed coating, flocking, electrostatic spraying, injection and compression molding.
5. Mechanical properties are not affected by long-term exposure in air at 450° F.
6. Selected for molding primarily where chemical resistance and high temperature properties are of prime importance.

Glass-Bonded Mica

1. This combination of a mica filler and glass binder provides high dimensional stability and can operate at extremely high or low temperature.
2. It is moisture and radiation resistant and will not outgas in vacuum.
3. Is outstanding for electrical and high heat (1,000°F plus) applications.

Methylpentene

1. Excellent transparency, lowest specific gravity of any commerial plastic (0.83).
2. Has good chemical resistance and low water absorbtion.
3. Can be used continuously at 275° F.

Chlorinated Polyether

1. Has strong resistance to both organic and inorganic agents, with the exception of strong oxidizing agents. This property gives chlorinated polyether a distinct advantage over many other chemical and processing application materials.
2. High thermal and chemical stability.

Polyimides (Thermoplastic)

1. They become rubbery rather than melt at approximately 580°F.
2. They can be molded by exceeding the glass transition temperature for a sufficient length of time.
3. Molded parts have good machineability and low brittleness.
4. They do not need postcure to develop high-temperature properties.
5. They can be reclaimed and used as regrind.
6. They can be cast into films from solution using conventional casting machines and directly coated onto substrates such as copper for circuit-board components.
7. They can be compression molded at approximately 640°F and 3,000 to 5,000 psi.

Polyimides (Thermoset)

1. Moldings and laminates can resist 500°F and up for several thousand hours; inherently resistant to combustion. Moldings can be used intermittently at temperatures as high as 900°F.
2. Polyimide adhesives maintain useful properties for well over 12,000 hours at 500°F, 9,000 hours at 575°F, 500 hours at 650°F, 100 hours at 700°F, and 25 hours at 750°.
3. Polyimide film has good mechanical properties through a range from liquid helium temperature to 1,100°F.
4. Films have high tensile and impact strength and high resistance to tear initiation.
5. Polyimides exhibit high dielectric strengths, low electric constants, good corona resistance, and low dissipation factors over a wide range of temperature and humidity conditions.

6. Polyimides are unaffected by exposure to dilute acids, aromatic and aliphatic hydrocarbons, esters, ethers, alcohols, Freon, hydraulic fluid, JP-4 fuel, and kerosene. They are attacked by dilute alkali and concentrated inorganic acids.

Phenolic

1. Good heat and chemical resistance, good surface hardness and dimensional stability, good electrical insulating properties, comparatively low cost.
2. Used extensively in compression and transfer molding. Also in adhesives, coatings, and laminating resins.
3. Products include automotive coil tops, distributor caps, and rotors. Also used in handles, knobs, telephones, circuit breakers, and adhesives for plywood.
4. Inherently brittle by itself, but addition of fillers and carriers results in strong, rigid molded parts.
5. Molding compounds are limited in color to dark shades of brown, green, blue, red, purple, and black.

Phenol-Furfural

1. A phenolic with improved chemical resistance.
2. Coatings formulated with this resin have high gloss and good impact strength.
3. Especially useful in molding of large parts because of long duration of flow.
4. Natural color of resin is black.

Urea

1. Clear, water-white resin. Cannot be used as clear casting material because of poor aging properties.
2. Combines with alpha cellulose for molding applications. Also used in adhesives for wood, coatings for paper, and crush-proofing impregnants for textiles.
3. Unlimited color range, glossy surface.

Melamine

1. Like urea, a member of the amino family of resins.
2. Properties are very similar to urea except that melamine offers improvement in heat resistance and moisture resistance.
3. Used extensively in the production of plastic dinnerware and decorative table tops and counter tops. Has harder surface than most plastic materials.

Polyester (Figure 16-2)

1. Strong, versatile materials with good weathering properties.
2. Chemical and solvent resistant, good colorability.
3. Used primarily with fillers and glass reinforcements in low-pressure molding and hand layups.
4. Products include boat hulls, automobile bodies, and translucent roof panels. Clear casting materials are used for imbedding scientific and decorative items.

Alkyd

1. Member of polyester family.
2. Good resistance to heat, dimensional stability, and moisture resistance.
3. Attacked by strong acids and bases.
4. Used primarily in paints and enamels for walls, automobiles, and appliances.
5. Used in compression and transfer molding.

Allyls

1. Retain electrical properties in high heat and high humidity.
2. Diallyl phthalate (DAP) is most widely used of allyl family of resins.
3. DAP is resistant to most chemicals and solvents. Also has good temperature resistance and very low moisture absorbtion.
4. Diallyl isophthalate (DAIP) is similar to DAP but is superior in impact strength and chemical resistance.

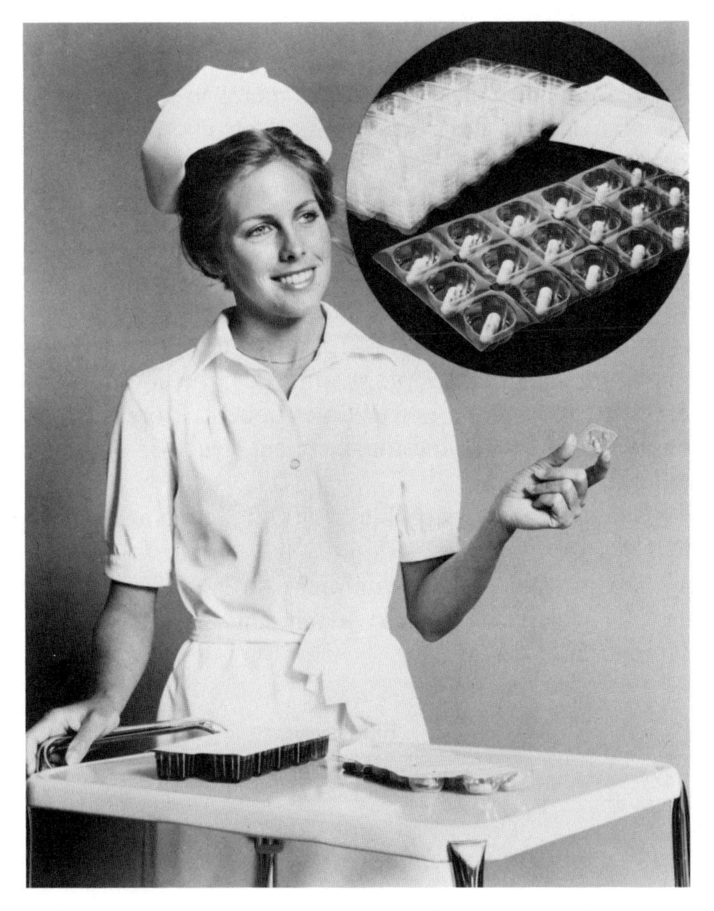

Figure 16-2. Unit-dose packaging trays made of extruded copolyester. *(Courtesy of Eastman Chemical Products, Inc.)*

5. Processing is by compression and transfer molding.
6. Allyl products include parts for trucks, trailers, aircraft, and spacecraft.

Epoxy

1. One of the most widely used plastic materials. Features good toughness, excellent chemical resistance, good heat resistance, low moisture absorption, and good adhesion to similar or dissimilar surfaces.

2. Processing includes molding, high- and low-pressure laminating, casting, coating, and adhesive bonding.
3. Products include tools for shaping metal in automobile and aircraft plants, adhesives, boat hulls, electronic parts, and tool fixtures for industry.

Silicone

1. Most diversified of all plastic materials. Can be used as molding compounds, laminating resins, release agents, adhesives, sealants, and potting compounds. End products can be liquids, semiliquids, greases, rigid laminates, or flexible rubbers. Property changes are brought about by varying the chemical structure and by addition of fillers and carriers.
2. Most silicones are serviceable to 500°F; some are serviceable to 700°F.
3. Does not provide good interlaminar strength like epoxies. Care must be exercised when machining silicones to prevent delamination.

Fillers

Filler	Function
Aluminum oxide	Used in adhesives for thermal conduction.
Aluminum powder	Adds hardness and acts as a heat sink to reduce exothermic heat. Good heat conductor. Used with epoxy to fill holes in aluminum.
Asbestos	Increases heat resistance and flame resistance. Decreases shrinkage and water absorbtion. Contributes to flow uniformity in premix molding compounds. Increases specific gravity. Has poor color properties.
Boron fibers	Provides high modulus.
Calcium carbonate, talc, clay, etc.	Lowers exothermic heat, reduces thermal expansion and contraction, increases physical strength, extends material mass.
Carbon fibers	Same as for graphite fibers.
Cotton flock	Adds impact strength.
Fiberglass	Provides mechanical strength.
Graphite fibers	Used in ablative, heat-resistant, and high-modulus applications.
High silica	Used in ablative and heat-resistant applications.
Macerated fabrics	Adds high impact strength and lowers mold shrinkage. Has high water absorbtion, high bulk factor, and poor electrical properties.

Filler	Function
Mica	Has excellent electrical properties. Also used in heat-resistant formulations. Reduces mechanical strength and increases specific gravity.
Microballoons	Used in syntactic foams and other weight reducing applications. Five-mil-diameter spheres are used in some adhesives to insure a 5-mil glue line.
Sand	Used in coatings to make decks and other surfaces skid proof and slip proof.
Shell flour	Adds high luster and gloss.
Silica powder (fumed)	Thickening agent for resins.
Silver	Used in epoxy adhesive formulations (loaded to approximately 80%) for electrical conductivity, especially where hot solder is impractical.
Wood flour	Adds moldability and high surface gloss. Reduces cost.

Mechanical and Physical Properties of Plastics

Tensile Strength		Compressive Strength	
Polymer	psi	Polymer	psi
Nylon	10–15,000	Epoxy	35,000
Polysulfone	10,500	Phenolic	34,000
Polyphenylene oxide	9,800	Melamine	30,000
Polycarbonate	9,500	Cellulose acetate	26,000
Acetal	9,000	Diallyl phthalate	22,000
Acrylic	8–10,000	Polyurethane	20,000
Phenolic (cast)	8–10,000	Polyphenylene oxide	16,000
Phenoxy	8,500	Acrylic	15,000
Melamine	8,000	Silicone	15,000
Polyester (cast)	8,000	Polystyrene	14,000
Cellulose acetate	5–8,000	Polysulfone	14,000
ABS	6–8,000	Nylon	13,500
Polystyrene	6–8,000	Acetal	13,000
Diallyl phthalate	6,000	Polycarbonate	13,000
Polyurethane	6,000	Polycarbonate	13,000
Polyethylene, H.D.	5,000	Phenoxy	11,000
Polypropylene	5,000	Polyvinyl chloride (rigid)	11,000
Fluorocarbons	5,000	ABS	10,500
Polyvinyl chloride (rigid)	5,000	Ionomer	8,000
Pollyallomer	3,800	Polypropylene	7,000
Polyethylene, L.D.	1,800	Polyallomer	4,200
		Polyethylene, H.D.	3,000
		Polyethylene, L.D.	2,200
		Fluorocarbons	1,700

Impact Strength		Flexural Strength	
Polymer	ft lb	Polymer	psi
Polycarbonate	8–13	Phenylene oxide	15,000
Polyvinyl chloride (rigid)	5–10	Acetal	14,000
Polyurethane (rigid)	5	Polycarbonate	13,000
ABS	4–6	Acrylic	8–18,000
Ionomer	3	Polystyrene	9–14,000
Pollyallomer	3	Polyvinyl chloride (rigid)	10–15,000
Polyethylene, H.D.	3	Polyethylene	
Fluorocarbons	3	low density	—
Polyphenylene oxide	4	high density	9,000
Cellulose acetate	2–5	Cellulose acetate	6–12,000
Polypropylene	2	Epoxy	
Polysulfone	2	cast	14–20,000
Phenoxy	1.5	glass laminates	40–100,000
Nylon	1.5	Phenolic	
Acetal	1.5	glass laminates	20–70,000
Epoxy	1.0	Polyester	
Polyester	0.8	glass laminates	30–90,000
Polystyrene	0.8		
Diallyl phythalate	0.5		
Acrylic	0.4		
Melamine	0.30		
Phenolic (cast)	0.30		

Specific Gravity

Polymer	Sp Gr	Fibers	Sp Gr
Methylpentene	0.83	Dacron (High Str.)	1.38
Polypropylene	0.89–0.90	Kevlar 49	1.45
Pollyallomer	0.89–0.90	Graphite (High Str.)	1.8–1.97
Polyethylene, L.D.	0.91–0.92	Asbestos	2.5
Polyethylene, H.D.	0.94–0.96	Fiberglass	2.54
Ionomer	0.93–0.96	High silica	2.6
Polyester	1.01–1.20	Quartz	2.6
ABS	1.04	Boron	2.65
Polystyrene	1.02–1.1		
Polyphenylene oxide	1.08		
Nylon	1.13–1.19	Metals	Sp Gr
Epoxy	1.11	Aluminum	2.5–2.7
Polyurethane	1.11–1.25	Steel	7.6–7.8
Vinyls	1.16–1.55		
Phenoxy	1.17		
Acrylic	1.17–1.20		
Polycarbonate	1.20		
Polysulfone	1.24		
Phenolic	1.25–1.30		

Polymer	Sp Gr	Fibers	Sp Gr
Cellulose acetate	1.2–1.34		
Polyvinyl chloride	1.15–1.35		
Diallyl phthalate	1.34–1.78		
Acetal	1.40–1.42		
Melamine	1.48		
Fluorocarbons	2.12–2.20		

Continuous Resistance to Heat		Water Absorbtion (24-hour test)	
Polymer	Temperature (F°)	Polymer	% Absorbed
Polyimide	600–700	Fluorocarbons	0.00
Silicone	600	Polyethylene, H.D.	0.01
Polyamide-imide	500	Pollyallomer	0.01
Teflon (TFE)	500	Polypropylene	0.01
Diallyl phthalate	425	Polystyrene	0.03
Phenolic	400	Polyphenylene oxide	0.05
Epoxy	400	Epoxy	0.10
Polyphenylene oxide	375	Silicone	0.11
Polypropylene	300	Phenoxy	0.13
Nylon	300	Polycarbonate	0.14
Polysulfone	300	Polysulfone	0.21
Polycarbonate	250	Acetal	0.21
Polyester	250	Acrylic	0.28
Polyethylene, H.D.	225	Ionomer	0.30
Acetal	220	Melamine	0.34
Ionomer	220	ABS	0.34
Melamine	220	Diallyl phthalate	0.35
Pollyallomer	210	Polyvinyl chloride	0.40
ABS	200	Polyester	0.50
Polyethylene, L.D.	200	Phenolic	0.60
Urethane	200	Polyurethane	0.75
Acrylic	180	Nylon	1.45
Cellulose acetate	180	Cellulose acetate	3.85
Polyvinyl chloride	175		
Polystyrene	170		
Phenoxy	170		

Self-Extinguishing	Clear-Molding Compounds
Melamine	Acrylic*
Urea	Phenoxy*
Polysulfone	Methylpentene*
Nylon	Polystyrene*
Vinyls	Cellulose acetate
Polycarbonate	Epoxy
Phenolics	Ionomer
	Polycarbonate
	Polyester

*Crystal clear.

Suitable for Outdoors	Good Resistance to Organic Solvents
Cellulose acetate butyrate	Fluorocarbons
Cellulose propionate	Nylon
Urethane	Urethane
Polyvinyl chloride	Ionomer
Epoxy	Epoxy
Polyester	Phenolic
Acrylic	Chlorinated polyether
Silicone rubber	Polyolefins
	Melamine
Good Abrasion Resistance	Acetal
Urethane	Polysulfone
Rigid vinyl	
Acetal	**Good Hinge Properties**
Nylon	Pollyallomer
ABS	Polypropylene

Measurement

Metric Measurement	U.S. Measurement
1 kilometer (km) = 1000 m	12 in. = 1 ft
1 meter (m) = 100 cm	3 ft = 1 yd
1 centimenter (cm) = 10 mm	1 cu ft = 1728 cu in.
1 liter (l) = 1000 ml	1 ton = 2000 lg
Conversion Factors for Length	
1 inch (in.) = 2.54 cm	1 centimeter = 0.3937 in.
1 foot (ft) = 30.48 cm	1 meter = 3.281 ft
1 mile (mi) = 1.609 km	1 kilometer = 0.6214 mi
Conversion Factors for Volume	
1 liter = 1.057 liquid qt	1 cubic foot = 28.32 liter
1 liquid quart = 0.9463 liter	1 cubic inch = 16.39 cc or ml
1 gallon = 3.785 liter	1 fluid ounce = 29.57 cc or ml
Conversion Factor for Mass	
1 pound = 453.6 grams	1 kilogram = 2.205 lb
1 ounce = 28.3 grams	1 gram = 15.43 grains

LENGTH-MEASURING DEVICES

While inspectors use a variety of measuring devices to check dimensions of finished parts, lab technicians normally depend upon three types for in-process work. These instruments are micrometers, vernier or dial calipers, and steel rule scales.

Micrometers

Micrometers measure outside dimensions. The 1-in. micrometer, which is the most widely used, consists of a rotating thimble that moves across a graduated sleeve or barrel. A spindle that is attached to the thimble moves back and forth with the movement of the thimble. The reading is taken when the spindle comes to a stop against the part being measured. Each mark or graduation on the sleeve is 0.025 in. Four of these marks equal 1/10 in. (0.10 in.). Each 1/10 in. is identified with a number, starting with No. 1. To arrive at the correct reading, note the largest number showing on the sleeve, then add 0.025 in. to it for each mark visible beyond the number. To this total, add the reading of the thimble. There are 25 marks on the thimble, each one representing 1/1000 in. (0.001 in.). See Figure 16-3 for a sample reading.

When necessary, readings can be taken to the nearest 1/10,000 in. (0.0001 in.) by using the vernier scale located at the top of the sleeve. A mark on the vernier will coincide with a mark on the thimble. The

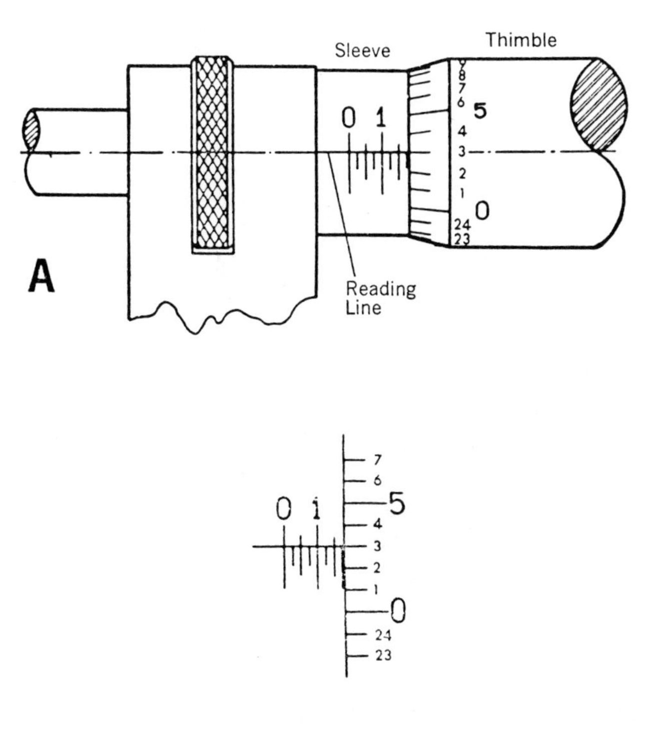

Figure 16-3. Micrometer reading. *(Courtesy of The L. S. Starrett Company)*

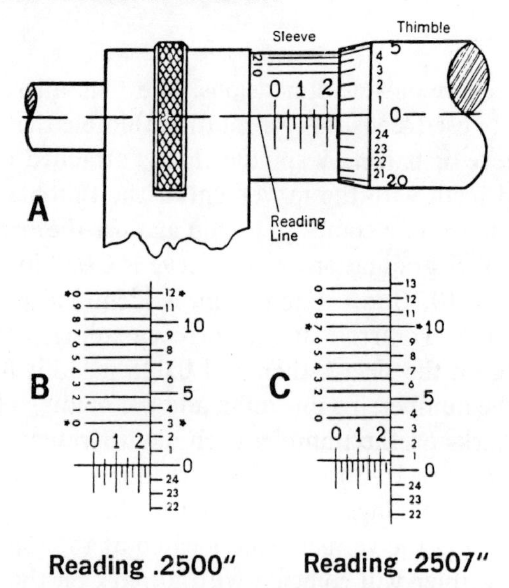

Figure 16-4. Micrometer readings, using the vernier scale. *(Courtesy of The L. S. Starrett Company)*

reading is taken at this point. The reading is in ten thousandths and is added to the subtotal. See Figure 16-4 for vernier readings.

Vernier Calipers

Vernier calipers are graduated in a manner similar to micrometers, but while micrometers can only be used to measure outside dimensions, vernier calipers can measure outside and inside dimensions and depth.

Instead of having a round, rotating member like micrometers, vernier calipers are flat and have a sliding member that performs the same function as the micrometer thimble. The slide also contains a vernier scale, which measures to the nearest 1/1000 in. (0.001 in.).

Step-by-step operation of a vernier caliper is as follows:

1. Note the largest inch number and ⅒-in. number left of zero. (Each number between inch marks is ⅒ in.).
2. Add 0.025 in. for each mark between the ⅒ in.-number and zero.
3. There are 25 numbers on the vernier scale, each representing

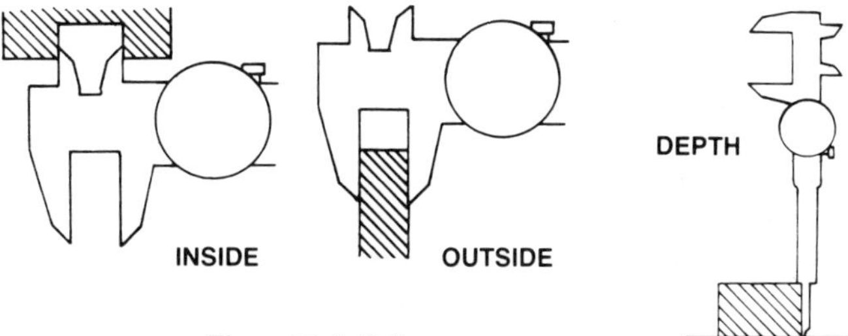

Figure 16-5. Caliper measurement.

0.001 in. One of these marks will coincide with a mark on the stationary rule. The reading on the vernier scale at this point is added to the subtotal to arrive at the final reading.

Dial Calipers (Figure 16-5)

Dial calipers are almost the same as vernier calipers. The only difference is that, instead of a vernier scale, the dial caliper has a 360-degree dial graduated in 1/1000-in. subdivisions. Hundredths and thousandths are read directly off the dial.

Steel Scales

Most steel scales used in plastics labs are ½ in. wide and either 6 in. or 12 in. long. They are thin and flexible and very handy for taking quick, rough measurements. Six-inch scales are especially popular. Being so small and lightweight they can be easily tucked away in a pocket, ready for instant use at any time, in any area.

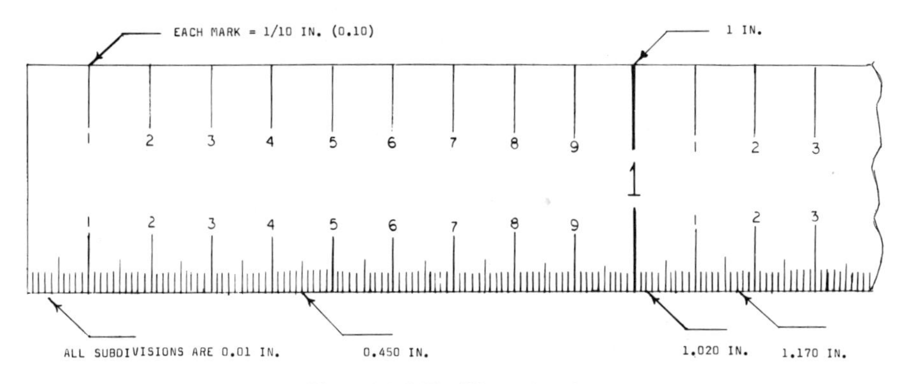

Figure 16-6. Flexible steel scale.

One side of most scales is graduated in 1/32-in. and 1/64-in. subdivisions. The other side, as noted in Figure 16-6, is graduated in 1/10-in. (0.10 in.) and 1/100-in. (0.100 in.) subdivisions.

DENSITY

Density is the ratio of weight to volume, normally expressed as pounds per cubic foot or pounds per cubic inch. Specific gravity (sp gr) is also an expression of density, but is specifically the ratio of weight in grams to volume in cubic centimeters (cc).

$$\text{sp gr} = \frac{\text{wt (grams)}}{\text{vol (cc)}}$$

All specific gravity computations are based upon the weight of water, which has a specific gravity of 1.0. This tells you that 1 cc of water weighs 1 gram. If 1 cc of another material weighed 2 grams, the specific gravity of the material would be 2.0. When two factors of the specific gravity formula are known, the third can be computed.

$$\text{wt(gram)} = \text{sp gr} \times \text{vol(cc)}$$

$$\text{vol(cc)} = \frac{\text{wt(grams)}}{\text{sp gr}}$$

$$1 \text{ cu in.} = 16.4 \text{ cc}$$

$$1 \text{ lb} = 454 \text{ grams}$$

One cubic foot of water weighs 62.4 lb.

$$\frac{\text{density(lb/cu ft)}}{62.4} = \text{sp gr}$$

$$\text{sp gr} \times 62.4 = \text{lb/cu ft}$$

$$\text{sp gr} \times 0.0361 = \text{lb/cu in.}$$

$$\text{lb/cu in.} \times 1728 = \text{lb/cu ft}$$

To convert a volumetric reading (ml) of a liquid to weight, multiply the reading by the specific gravity of the liquid.

Example: What is the weight of 100 ml of a liquid having a specific gravity of 1.5?

Answer: wt = 1.5 × 100
wt = 150 grams

To convert the weight of a liquid to volume (ml), divide the weight by the specific gravity of the liquid.

Example: What is the volume (ml) of 120 grams of a liquid that has a specific gravity of 1.2?

Answer: $$\text{vol} = \frac{120}{1.2}$$

$$\text{vol} = 100 \text{ ml}$$

Specific gravity of liquids can be measured by hydrometers, which are graduated, weighted tubes that sink part way in a liquid column. The tests are usually conducted in a graduate with readings taken at the meniscus line, which is the concave line formed at the top of the liquid column. The meniscus is always concave when the containing walls are wetted by the liquid. Mercury, the only metallic liquid, forms a convex meniscus because it does not wet the sides of the container.

Plastics Trade Names

Trade Name	Material	Manufacturer
Aclar	Fluoroplastic film	Allied Chemical Corp.
Acrylite	Acrylic molding material	American Cyanamid Co.
Araldite	Epoxy resins	Ciba Co., Inc.
Bakelite	Various resins	Union Carbide Corp.
Beetle	Urea molding compounds	American Cyanamid Co.
Catalin	Cast phenolics	Catalin Corp. of America
Celanese	Resins and films	Celanese Plastics Co.
Cycolac	ABS	Marbon Chemical Co.
Dacron	Polyester fiber	Du Pont
Dapon	Allylic resins	Food Machinery & Chemical Corp.
Delrin	Acetal resin	Du Pont
Durez	Molding compounds	Hooker/Durez
Epon	Epoxy resins	Shell Chemical Corp.
Epotuf	Epoxy resins	Reichold Chemicals, Inc.
Epoxical	Resins for tooling	U.S. Gypsum
Fiberfil	Glass-filled injection molding compounds	Fiberfil, Inc.
Formica	Decorative laminates	Formica Corp.
Geon	Vinyl resins	B. F. Goodrich Co.
Glaskyd	Glass-reinforced alkyd	American Cyanamid Co.
Halex	Fluorocarbon molding compound	Allied Chemical Corp.
Hetron	Fire-retardant polyester resin	Hooker Chemical Corp.
Hysol	Epoxy adhesives	Hysol Corp.
Kel-F	Chlorotrifluoroethylene molding resins	Minnesota Mining Co.

Plastics Trade Names (cont'd)

Trade Name	Material	Manufacturer
Lexan	Polycarbonate resin	General Electric Co.
Lucite	Acrylic resin	Du Pont
Lustran	ABS resin	Monsanto Co.
Lustrex	Polystyrene molding and extrusion compounds	Monsanto Co.
Marvinol	Vinyl chloride resins	Uniroyal, Inc.
Melmac	Melamine molding compounds	American Cyanamid Co.
Micarta	Thermosetting laminates	Westinghouse Corp.
Mycalex	Glass-bonded mica	Mycalex Corp.
Mylar	Polyester film	Du Pont
Orlon	Acrylic fiber	Du Pont
Opalon	Vinyl chloride resins	Monsanto Co.
Paraplex	Polyester resins	Rohm and Haas Co.
Pelaspan	Expandable polystyrene bead	Dow Chemical Co.
Plaskon	Various resins	Allied Chemical Corp.
Plexiglass	Acrylic sheets and molding materials	Rohm and Haas Co.
Polypenco	Nylon rods, tubes, filaments, film, and sheets	Polymer Corp.
Poly-pro	Polypropylene	Gulf Oil Corp.
Royalite	ABS sheets	Uniroyal, Inc.
Saran	Vinylidene chloride copolymers	Dow Chemical Co.
Silastic	Silicon rubber	Dow Corning Co.
Styrafoam	Expanded polystyrene	Dow Chemical Co.
Surlyn	Ionomer resins	Du Pont
Tedlar	Polyvinyl fluoride film	Du Pont
Teflon	Tetrafluoroethylene (TFE) and fluorinated ethylpropyline (FEP)	Du Pont
Tenite	Cellulosics and polypropylene	Eastman Chemical Co.
Textolite	High-pressure, decorative, and industrial laminates	General Electric Co.
Ultrathene	Ethylene-vinyl acetate	U.S. Industrail Chemicals Co.
Versamid	Polyamide resins	General Mills, Inc.
Zytel	Nylon molding powder	Du Pont

Abbreviations

ASTM—American Society of Testing and Materials
FRP—Fiberglass-reinforced plastics
PBW—Parts by weight
PHR—Parts per hundred resin
Pre-preg—Preimpregnated
RP—Reinforced plastics
RP/C—Reinforced plastics composites
RTP—Reinforced thermoplastics

SPE—Society of Plastics Engineers
Spex—Specifications
SPI—Society of Plastics Industry

Shop Jargon

Advancing resin—Catalyzed resin starting to cure
Ears—Folds made in a vacuum bag to take up slack
Goup or **gupe**—Resin or adhesive
Gunk—Thick, viscous resin mix
Kicked-over resin—Resin that has solidified
Shoot—Injection or spraying of liquid resin systems
Skins—Facings on honeycomb panels
Speegee—Specific gravity
Splash—Casting plaster over a shape to be reproduced
Staged—Refers to B-stage

Temperature Conversion

C°	F°	C°	F°
0	32	90	194
5	41	95	203
10	50	100	212
15	59	105	221
20	68	110	230
25	77	115	239
30	86	120	248
35	95	125	257
40	104	130	266
45	113	135	275
50	122	140	284
55	131	145	293
60	140	150	302
65	149	155	311
70	158	160	320
75	167	165	329
80	176	170	338
85	185	175	347

Conversion Formulas

$$C^{\circ} \times \frac{9}{5} + 32 = F^{\circ} \qquad \frac{5}{9}(F^{\circ} - 32) = C^{\circ}$$

The plastics industry normally uses the Fahrenheit scale for temperature callouts. The chemical industry uses centigrade (Celsius) degrees.

Areas and Circumferences of Circles

Diam.	Circum.	Area	Diam.	Circum.	Area	Diam.	Circum.	Area
7.	21.991	38.485	⅜	32.594	84.541	15.	47.124	176.71
⅛	22.384	39.871	½	32.987	86.590	¼	47.905	182.65
¼	22.776	41.282	⅝	33.379	88.664	½	48.695	188.69
⅜	23.169	42.718	¾	33.772	90.763	¾	49.480	194.83
½	23.562	44.179	⅞	34.165	92.886			
⅝	23.955	45.664				16.	50.265	201.06
¾	24.347	47.173	11.	34.558	95.033	¼	51.051	207.39
⅞	24.740	48.707	⅛	34.950	97.205	½	51.836	213.82
			¼	35.343	99.402	¾	52.622	220.35
8.	25.133	50.265	⅜	35.736	101.62			
⅛	25.525	51.849	½	36.128	103.87	17.	53.407	226.98
¼	25.918	53.456	⅝	36.521	106.14	¼	54.192	233.71
⅜	26.311	55.088	¾	36.914	108.43	½	54.978	240.53
½	26.704	56.745	⅞	37.306	110.75	¾	55.763	247.45
⅝	27.096	58.426						
¾	27.489	60.132	12.	37.699	113.10	18.	56.549	254.47
⅞	27.882	61.862	¼	38.485	117.86	¼	57.334	261.59
			½	39.270	122.72	½	58.119	268.80
9.	28.274	63.617	¾	40.055	127.68	¾	58.905	276.12
⅛	28.667	65.397						
¼	29.060	67.201	13.	40.841	132.73	19.	59.690	283.53
⅜	29.452	69.029	¼	41.626	137.89	¼	60.476	291.04
½	29.845	70.882	½	42.412	143.14	½	61.261	298.65
⅝	30.238	72.760	¾	43.197	148.49	¾	62.046	306.35
¾	30.631	74.662						
⅞	31.023	76.589	14.	43.982	153.94	20.	62.832	314.16
			¼	44.768	159.48	¼	63.617	322.06
10.	31.416	78.540	½	45.553	165.13	½	64.403	330.06
⅛	31.809	80.516	¾	46.338	170.87	¾	65.188	338.16
¼	32.201	82.516						

$$A = \pi R^2 \qquad A = \frac{\pi D^2}{0.785} \qquad A = 0.785D^2$$

$$C = \pi D \qquad \pi = 3.1416$$

Conversion of Fractions to Decimals

Fraction	Decimal	Fraction	Decimal
1/64	0.0156	33/64	0.5156
1/32	0.0312	17/32	0.5312
3/64	0.0468	35/64	0.5468
1/16	0.0625	9/16	0.5625
5/64	0.0781	37/64	0.5781
3/32	0.0937	19/32	0.5937
7/64	0.1093	—	—
1/8	0.125	—	—
9/64	0.1406	39/64	0.6093
5/32	0.1562	5/8	0.625
11/64	0.1718	41/64	0.6406
3/16	0.1875	21/32	0.6562
13/64	0.2031	43/64	0.6718
7/32	0.2187	11/16	0.6875
15/64	0.2343	45/64	0.7031
1/4	0.250	23/32	0.7187
17/64	0.2656	47/64	0.7343
9/32	0.2812	3/4	0.75
19/64	0.2968	49/64	0.7656
5/16	0.3125	25/32	0.7812
21/64	0.3281	51/64	0.7968
11/32	0.3437	13/16	0.8125
23/64	0.3593	53/64	0.8281
3/8	0.375	27/32	0.8437
25/64	0.3906	55/64	0.8593
13/32	0.4062	7/8	0.875
27/64	0.4218	57/64	0.8906
7/16	0.4375	29/32	0.9062
29/64	0.4531	59/64	0.9218
15/32	0.4687	15/16	0.9375
31/64	0.4843	61/64	0.9531
1/2	0.50	31/32	0.9687
		63/64	0.9843
		1	1.000

Tap/Drill Sizes

Tap Size	Drill Size	Tap Size	Drill Size
0–80	3/64	1/4–20	7
1–64	53	1/4–28	3
1–72	53	5/16–18	F
2–56	50	5/16–24	1
2–64	50	3/8–16	5/16
3–48	47	3/8–24	Q
3–56	45	7/16–14	U
4–40	43	7/16–20	25/64
4–48	42	1/2–13	27/64
5–40	38	1/2–20	29/64
5–44	37	9/16–12	31/64
6–32	36	9/16–18	33/64
6–40	33	5/8–11	17/32
8–32	29	5/8–18	37/64
8–36	29	11/16–11	19/32
10–24	25	11/16–16	5/8
10–32	21	3/4–10	21/32
12–24	16	3/4–16	11/16
12–28	14	7/8–9	49/64
14–20	10	7/8–14	13/16
14–24	7	1–8	7/8

REFERENCES

"Plastics Properties Chart," *Modern Plastics Encyclopedia, 1967–1977*. McGraw-Hill, New York, 1976.

Baird, R. J. *Industrial Plastics*. Goodheart-Wilcox Company, Inc.

Milby, R. V. *Plastics Technology*. McGraw-Hill, New York, 1973.

Simonds, Church. *A Precise Guide to Plastics*. Van Nostrand Reinhold Co., New York, 1963.

Glossary

Ablative plastics. Plastics composites used as heat shields on aerospace vehicles. The intense heat encountered erodes and chars the top layers. The charred layer and some cooling effects from evaporation insulate the inner areas against further penetration of the intense heat.

Adherends. Two bodies held together by an adhesive. **Substrate** has a similar, though expanded, meaning. It is any body that something is bonded to or otherwise attached to. **Faying surface** is the bonding surface of an adherend or substrate.

Adhesive failure. This undesirable type of failure occurs when a lap shear specimen fails by leaving adhesive on one adherend and none on the corresponding area of the other adherend. **Cohesive failure** is when failure occurs within the adhesive, leaving adhesive on both adherends within the same contact area. When both shear areas are completely covered, it is called a 100% cohesive failure—the optimum condition.

Advancing. In-progress chemical reaction that results in curing the resin.

Ambient temperature. Temperature of the surrounding area.

Amorphous. Having no regular shape; shapeless.

Annealing. The process of heating a material just below its heat distortion point to relieve stresses.

A-stage. Thermoset resin in the uncured state. **B-stage** describes a partially cured state and **C-stage** is a fully cured state.

Autoclave. Pressure vessel used to cure assemblies that require more than vacuum pressure. Autoclaves capable of 50–100 psi are most common.

Ball mill. Small machine with rubber-faced rollers that is used to rotate cans or other containers for the purpose of dissolving solid resins in organic solvents. The resin is cut into small pieces, or may already be in a powdered or granular form. Ceramic balls are added to the ingredients inside the container to provide friction and movement.

Bias cut. Cutting material 45 degrees from the weave pattern.

Binder. Resin that holds together filler or carriers in a pre-preg or molding compound. **Carrier** is the fabric portion of a composite. Fiberglass is the most commonly used fabric for composites.

Blocking. The sticking together of layers of material during storage. A separating film of polyethylene can be used to prevent this undesirable adhesion.

Blowing. The rising and expanding of a material during a foaming operation.

Boss. Protrusion designed for added strength or attachment of fasteners.

Breathing. Degassing of materials (for example, phenolics) that give off gasses during compression molding. After the mold has been closed for several seconds, it is opened a crack for several more seconds to allow trapped gasses to escape. With some materials the operation is repeated.

Bulk factor. Ratio between a cured and uncured molding compound for compression molding. Light, fluffy, shredded materials have a low uncured density and, therefore, a high bulk factor. The question of concern is whether high-bulk-factor material fits in the mold cavity.

Calibrate. To check and make accurate measurements using measuring devices such as calipers, recorders, scales, and balances. Checking is carried out at periodic intervals. A due date is attached to the instrument, and it cannot be used beyond this date until recalibrated.

Calipers. Measuring devices, such as micrometers and vernier calipers, having a graduated scale for exact measurement.

Carrier. See Binder.

Catalyst. Chemical that agitates or causes a chemical reaction but does not itself become part of the new compound. An example would be the peroxides used to cure polyesters. By contrast, most epoxy curing agents are 100% reactive and become part of the chemical change.

Caul plates. Metal plates between which laminates are pressed.

Charge. The weight of material used to make one molded part.

Cold Flow See Creep.

Configuration. Shape.

Concave surface. Surface curved inward, such as the inner surface of a salad bowl.

Contact cement. Mixture of elastomeric rubber and organic solvents that cure upon evaporation of the solvent. A coating is applied to both bonding surfaces. The surfaces are then joined together when an aggressive tack develops within several minutes).

Contact pressure. Application of little or no pressure on a lamination or bonded assembly.

Convex surface. Surface curved outward, such as the surface of a sphere.

Copolymerization. Material made by combining two different monomer systems.

Crazing. Development of fine cracks on a plastic surface.

Creep. Deformation or change of dimensions of a material under a constant load over a period of time, after the initial deformation when the load is first applied. The occurrence at room temperature is sometimes called **cold flow.**

Cryogenic. Producing very low temperatures.

Curing. Converting a resin from a flowable mass to a tack-free solid by addition of a curing agent.

Daylight opening. The clearance between platens when the press is fully opened. The opening must be large enough to allow a part to be ejected when the mold is in a fully opened position.

Delamination. The separation between plies of a fully cured laminate.

Desiccant. Chemical used for drying purposes because of its affinity for water. A **desiccator** is a heat-resistant glass container having an air-tight lid and lower compartment that houses a desiccant. Hot test specimens are placed in a desiccator to prevent moisture pickup before weigh back.

Dielectric. A nonconductor of electricity. Plastic materials are dielectrics.

Diluent. A modifier added to a resin to lower the viscosity (i.e., to thin it down). In the case of epoxies, most diluents used are 100% reactive. This means that they react chemically to become part of the cured compound.

Double-back tape. Tape having adhesive and protective covering on both sides. The covering is removed when the tape is ready for use.

Drape. The bending or droop in a pre-preg. Drape is tailored to the specific application. A worker making a hand layup on sharp bends wants plenty of drape, but the press laminator does not need any drape at all.

Elastic limit. The extent to which a material can be stretched or deformed before taking on a **permanent set.** Permanent set occurs when a material that has been stressed does not recover its original dimensions, as when a 12-in. piece of rubber that has been stretched becomes 13 in. long when relaxed.

Elastic memory. The ability of a thermoplastic to return to its original shape when exposed to heat beyond its heat distortion point. A flat sheet that has been thermoformed to a new shape reverts to a flat sheet if sufficiently heated.

Elastomer. A material that can be stretched to at least twice its original length.

Exotherm. The heat created by some curing agents when reacting with resins. The highest temperature occurs in the center of the mix. Some mixes can become smoking hot, such as can occur when DTA. (diethylene triamine) is mixed with an epoxy in large volumes. Peroxides added in high proportions to polyesters containing cobalt naphthenate can actually catch fire. These superactive combinations are used primarily as binders. (DTA. is also used in adhesives.) They should not be used as casting materials except on the low end of the proportion range, and then only in shallow pours. Some materials such as urethanes, RTVs and epoxy-polymide mixes give off very little or no exothermic heat and can be potted in large volumes. **Endothermic** refers to the absorption of heat in a chemical reaction.

Extenders. Inert fillers added to resins for the purpose of increasing the volume of the resin mix.

Faying surface. See **Adherends.**

Flash. Cured excess resin squeezed from a molding or laminate.

Gel. The first stage of curing when a resin has thickened to a near solid.

Gel coat. The first coat applied to a mold in a hand layup operation. It is usually formulated to add characteristics like surface hardness and cosmetic attractiveness to finished products such as boat hulls, and automobile bodies. The gel coat is allowed to advance to a near-cured condition before application of the laminate resin and buildup plies. Gel coats are used only on those laminates requiring a special surface. Almost all are polyesters.

Glue line. The adhesive layer between two adherends.

Hygroscopic. Describes a material that absorbs moisture.

Inhibitor. A material added to a molding compound or other resin mix to prevent premature advancement.

Insert. Usually a metal inclusion that can be either set in the mold cavity before molding or press-fitted in place later.

Jig. Tool for holding parts together during a fabrication operation. Also known as a **fixture.**

Kick over. Shop jargon describing the curing of a thermoset resin to the solid state. **Set up** has same meaning.

Lagging. Process involving the wrapping of resin-impregnated tape around a cylindrical mandrel and applying pressure by shrink tape. Prestretched Tedlar is the most commonly used shrink tape. It shrinks upon application of heat.

Laminate. Bonding together of two or more plies of impregnated fabric. Fabrication can be by press, vacuum bag, or contact pressure alone.

Loading well. The top area of a mold cavity, the size of which is dictated by the bulk factor of the molding compound. High-bulk-factor materials require deeper wells than do low-bulk-factor materials (compression molding).

Layup. Placing of reinforcing fabric, such as fiberglass, and catalyzed resin on a mold. Cure is by vacuum pressure and heat or by contact pressure and no heat, depending on the resin system used. If a higher pressure is specified, the bagged layup is cured in an autoclave while still under vacuum.

Mat. Glass fibers, either chopped or continuous, formed into a felt. It can be used for hand layups or preshaped for low-pressure compression molding.

Meniscus. The top surface line of a liquid when enclosed in a container such as a graduate. This surface is always concave, except for mercury, which forms a convex meniscus. When one is measuring liquids, the meniscus should be at eye level.

Microspheres. Tiny hollow plastic spheres used to make lightweight plastic mixes such as syntactic foam: 5-mil-diameter spheres are occasionally used in adhesives to control glue-line thickness. When the adherends are pressed together the spheres will insure a glue-line thickness of 5 mil (1% spheres is the amount normally used).

Modulus. The ratio of stress to strain in a material when it is elastically distorted. A measurement of stiffness or resistance to bending. All computation for modulus is made within the elastic limit.

Muffle furnace. High temperature furnace used primarily to burn off cured resins for computation of resin content in laminates having carriers such as fiberglass.

Negative pressure. Refers to vacuum pressure.

Nominal. The closest approximate amount. Not exact; may vary somewhat.

Optimum condition. The ideal situation: the best possible condition.

Parison. The thermoplastic tube that is heated and blown to the configuration of the mold in a blow-molding operation. Plastic bottles are made in this manner.

Parting agents. These are materials, foremost of which are silicones and fluorocarbons, have no-stick properties. They are used on surfaces that must be kept free of contacting resins, foams, adhesives, etc. **Release agents** have the same meaning.

Periphery. The outer edges of an object.

Permanent set. See **Elastic limit.**

Plasticizer. A chemical added to vinyls to make them softer and more pliable. This combination is called a **Plastisol.** If a volatile thinner is added, it is called an **Organosol.** There are other materials that function similarly, such as the effect of versamids on epoxies.

Plasticize. To soften a material through the use of heat or by addition of a plasticizer.

Platens. The top and bottom contact surfaces of a press. Molding-press platens have facilities for mounting molds and are cored for heating and cooling. Heating is supplied mostly by steam or electricity, and cold water is used for cooling. One platen is movable, and the other is stationary.

Pop-off valve. Safety valve on pressure vessels that provides escape of excess pressure. The valve functions at a preset determination somewhere within the safety limit of the tank.

Pot life. The length of time that a material remains workable after addition of the curing agent. It is very important to know the pot life to prevent premature advancement of the resin. Pot life is measured at room temperature.

Preform. Compressed molding compound. Fits easily into the mold, simplifies preheating, and allows quicker closing of the press, which minimizes the possibility of case hardening. Preforming is especially effective when feeding multicavity molds. Another type of preform involves the matter formation of chopped glass fibers against a screen shaped to the contour of the mold. Pressure against the screen can be applied either by air or by water. Continuous glass fibers are also preshaped for mat molding. Both utilize polyester low-pressure systems.

Pre-preg. A resin-impregnated cloth, usually in a B-stage condition. Used for hand layups and press laminations. Filaments can also be pre-pregs.

Premix. On-site mixing of components to form a molding compound: most commonly associated with polyester A-stage mixes. See **A-stage.**

Primer. A coating applied to a bonding surface prior to the application of an adhesive to improve the quality of the bond. In some cases, a primer is applied and cured to protect the bonding surface during storage.

Release agents. See **Parting agents.**

Sandwich construction. Core materials such as honeycomb and foam sandwiched between two thin, high-strength facings or "skins."

Shrink fixture. Fixture upon which hot molded parts are placed so that they do not shrink during cool down.

Skins. Shop term applied to the outer ply of a laminate or honeycomb sandwich (one on each side).

Solvent bonding. Bonding together of two thermoplastic surfaces by application of an appropriate solvent. Ethylene dichloride is used for acrylics.

Step cure. Cures that are started at lower temperatures and gradually brought up to the cure temperature. This allows gasses to escape before solidification of the resin, as in the curing of phenolics.

Stress. The weight or pressure applied to a material, expressed in psi. **Strain** is the dimensional change of a material under stress, expressed in inches per inch.

Spex. Specification(s). A Process specification outlines materials, proportions, pot life, and cure cycles. The worksheet or shop order calls out the specific spex to be followed. Material spex give a more indepth study of the materials. Some process spex call out the materials by their material spex number, forcing the user to look further to learn what materials to use. This is a poor practice and is frowned upon by most technicians.

Substrate. See **Adherends.**

Syntactic foam. A lightweight mixture of thermosetting resin and microspheres (usually glass or phenolic). Areas on honeycomb structures are potted with syntactic foam to provide a solid foundation for installation of bushings and other hardware.

Tack primer. Sticky, tacky thermoset resin used to hold dry or nontack pre-preg together during a laminating or bonding operation.

Tear ply. The outer ply of a laminate that is designed for easy removal after cure, so as to provide a rough, bondable surface for a secondary bonding operation.

Thermoforming. Heating of a thermoplastic sheet beyond its heat distortion point, then forcing it over a preshaped mold so that when it cools down it retains the shape of the mold. Pressure on the sheet is usually applied by vacuum, but can also be applied by air pressure or mechanical means.

Thixotropic. The shop reference to this term describes a filled resin that has little or no movement when applied to a vertical plane. Powdered silica and other fillers are used as thickening agents.

Viscosity. The measurement of flow in a liquid. Water has a low viscosity. Molasses has a high viscosity. Measurement is in poises or centipoises.

Volatile. A liquid, such as an organic solvent, that continually releases vapors in air.

Volumetrically. Refers to the measuring of components by volume rather than weight; 150-cc and 250-cc poly beakers are common in a plastic lab for measuring of coating and paint components.

Water break. The appearance of water on a surface. On a clean, prepared surface, water will form a solid film. A discontinuous film indicates contamination and incomplete surface preparation.

Water-white. Having the clarity and appearance of water.

Wet bottle. Used in the packaging of adhesive components for on-site mixing. Low-proportion curing agents such as MEKP or Thermolite 12 are poured into the container then poured back out. The exact amount needed is then added to what remains in the container.

Index

Abbreviations for plastics, 230
Ablative materials, 104
Abrasion resistance, 213
ABS (Acrylonitrile butadiene styrene), 209
Acetal, 9, 213
Acrylic, 10, 81, 94, 95, 213
Adhesive bonding, 165
Alkyds, 23
Allylics, 24, 218
Amine curing agents, 98, 100
Aramid, 104, 109
Areas and circumferences of circles, 232
Asbestos, 128, 220
Autoclaves, 144, 235

Bag molding, vacuum, 135, 137
Bakelite, 1
Barcol Hardness Test, 205
Bell jar, 156, 157
Benzoyl peroxide, 99, 102, 127
Binders, 103
Bleeder materials, 137, 142
Blow molding, 71
Bonding techniques, 167
Boron fibers, 104
Brookfield Viscometer, 197
B-Stage, 5
Bulk factor, 115, 236

Calendering, 94
Calipers, 226
Carbon fibers, 104
Carving of plastics, 94
Casting, 6, 11, 147, 151
Catalysts, 6, 102, 127, 134, 179
Cellulosics, 11
  acetate, 11, 207
  acetate-butyrate, 207
  ethyl, 12, 207
  fabrication, 86
  nitrate, 11, 208
  propionate, 208
Chemistry of plastics, 8
Chrome plating, molds, 111
Chromic acid surface treatment (aluminum), 169
Clamps, 174
Clear molding compounds, 223
Coating, 171
Composites, 103
Compression molding, 111
  high pressure, 114
  low pressure, 127
Compression test, 205
Contact cement, 152, 168
Contact molding, 137, 192
Continuous pultrusion, 150
Cooling fixtures, 129
Core coating, 144, 145
CTFE (chlorotrifluoroethylene), 13, 212
Cultured marble, 148
Cure shrinkage, 6
Curing agents, 98
  epoxy, 98–101
  polyester, 98, 102

Degassing, 117
Degradation, thermal, 6
Density, 228
Diallyl resins, 24, 218
Diethylene triamine, 98, 100
Diluents, reactive, 102
Dip coating, 93
Drape, 191
Durometers, 206

Edge lighting, 95
Elastic tape molding, 149
Electronic labs, plastics, 155
Elongation, 204
Embedding, polyester, 151
Encapsulation, 155
Engraving of plastics, 94

Epoxies, 25, 219
  adhesive, 165
  casting, 147
  curing agents, 98–101
  potting, 147, 160
  resins, 98–101
Exotherm, 6
Extrusion, 80

Fabrication aids, 174
Fiberglass technology, 104
Fibers, 103
  Aramid, 104, 109
  asbestos, 128, 220
  boron, 104
  fiberglass, 103, 104, 134, 148
  graphite, 104
  high silica, 104
  thermoplastic, 127
Fiber volume test, 197
Filament winding, 148
Fillers, 6, 127, 220
Flexural test, 204
Flow test, pre-pregs, 192
Fluorocarbons, 13, 212
Foam, rigid urethanes, 162
Fume hood, 178
Furniture, plastic, 147

Gates, 57, 60, 123
Gel coat, 134
Glass-bonded mica, 215
Glossary, 235
Good work habits in the lab, 176
Graphite fibers, 104
Guide pins, 40, 113

Hand layup, 134
Hardness tests, 205
  Barcol, 205
  Shore, 206
Heat resistance, 223
Heat sealing, 93, 139
Honeycomb construction, 142–146
Hot staking, 95

Impact test, Izod, 205
Indentation hardness test, 205
Inhibitors, polyester, 102
Injection molding, 34
  machines, 34
  mold design, 54
  molding cycles, 43
  molding problems, 69
  molding shrinkage, 45
  pellet geometry, 38
  preheating, predrying, 37
  pressure, 37, 40
  screw design, 48
Inserts, 120, 125
Ionomer, 214

Kapton, 140
Kel-F (chlorotrifluoroethylene), 13, 212
Kevlar, 104, 109
Knock-out pins, 74

Lab scales, 185
Laminating, 134–145
Lap shear test, 200
Latex, 172
Leak detectors, 141
Length measuring devices, 224

Materials guide, 207
Measurements, U.S., metric, 224
Mechanical and physical properties of plastics, 191–208
Melamine, 227
Methylpentene, 215
Mica, 215, 221
Microballoons, 221
Micrometers, 225
Modulus, 201
Molding compounds, 5, 113, 114
Molding problems
  high-pressure compression, 121
  injection, 69
  low-pressure compression, 131
Mold releases, 111
Molds
  compression, 111
  extrusion, 35
  fiberglass, 135
  injection, 54
  plaster, 135
  plastic, 135
  transfer, 123

Nitrocellulose, 1, 11, 208
Nylon, 14, 212

One stage sandwiches, honeycomb, 143
Outgas, 155, 157
Ovens, 188, 177, 179

Parison, 71
Pellet geometry, injection molding, 38
Peroxide curing agents, 98, 102
Phenolic, 27, 117, 217
Phenyl silane, 115
Physical tests for plastics, 191
Plaster molds, 135
Plastic repairs, 152–154
Plastics trade names, manufacturers, 230
Plastisol molding, 93
Platen presses, 182–185
Polyallomer, 215
Polyamide, 14, 212
Polycarbonate, 15, 212
Polychlorotrifluoroethylene (Kel-F), 13, 212
Polyester, 15, 218
  casting, 11
  filament winding, 148
  layup, 134
  molding, 127
Polyethylene, 16, 208
Polyimide, 216
Polyphenylene oxide, 17, 214
Polyphenylene sulfide, 215
Polypropylene, 17, 209
Polystyrene, 18, 209
Polysulfone, 18
Polytetrafluoroethylene (PTFE), 12, 211
Polyurethane, 19, 213
Polyvinyl alcohol, 21, 211
Polyvinyl chloride, 21, 210
Polyvinyl fluoride, 230
Pot life, 7
Potting, 147, 160
Preform molding, 129
Preforms, 115
Preheating, 115
Premix molding, 127
Pre-pregs, 137, 186
  composition, 191
  drape, 191–192
  impregnation, 186
  tack, 191
Presses
  calculations for pressure, 117, 184
  laminating, 182–185
  molding, 111–113
Primers, RTV, 166
Promotors (accelerators), 6, 102

Quality control, 180, 191
Quartz, 103

Reinforced thermosets, 103
Release agents, 111
Repairs, 152–154
Resin content tests, 193–196
  acid digestion, 194
  burnout, 194
  measurement, 195
Rotational molding, 95
Rubber, silicone
  casting, 147
  molds, 148

Safety precautions, 178
Saran (polyvinylidene chloride), 210
Screeding, 136
Shelf life, 7
Shop jargon, 179, 231
Shop life, 7
Shore Durometer Hardness Test, 206
Shrinkage, 6
Sodium etchant, Teflon, 170
Specific gravity, 222, 228
Spray molding, 150
Surface preparation for bonding, 168
Syntactic foam, 221

Tape wrapping, 149
Tap sizes, 234
Tedlar, 149
Temperature conversion, C° to F°, 231
Tensile shear test, 201
Tensile test, 200
Thermal conductivity, 220
Thermoforming, 97
Thermoset labs, 181
Thermoset material forms, 5
Thixotropic fillers, 220, 221
Tooling, plastic, 135
Toxicity, 98
Transfer molding, 123

Traps, resin, 156
Triethylene tetramine, 98

Undercut, mold, 136
Urea, 32, 217
Urethanes, 213

Vacuum bag films, 139
Vacuum bag molding, 137
Vacuum bag sealants, 139
Vacuum bag valves, 139
Vacuum encapsulation, 155–159
Vacuum impregnation, 159
Vacuum pumps, 135

Venting, 65
Vernier calipers, 226
Vessels, pressure, 159
Vinyls, 20
Viscosity, 197
Void content test, 197

Water absorption, 223
Water break, 168
Weathering properties of plastics, 224
Welding, plastics, 94

Zinc chromate paste, 139
Zinc stearate, 128